Applied Photochromic Polymer Systems

To Angela

Applied Photochromic Polymer Systems

Edited by

C.B. McARDLE
Senior Chemist
Chemical and Materials Science Department
Loctite (Ireland) Limited
Dublin

Springer Science+Business Media, LLC

© 1992 Springer Science+Business Media New York
Originally published by Blackie & Son Ltd in 1992
Softcover reprint of the hardcover 1st edition 1992

First published 1992

British Library Cataloguing in Publication Data

Applied photochromic polymer systems.
I. McArdle, C.B.
668.4

ISBN 978-94-010-5356-3

Library of Congress Cataloging-in-Publication Data

Applied photochromic polymer systems / edited by C.B. McArdle.
    p.   cm.
Includes bibliographical references and index.
ISBN 978-94-010-5356-3   ISBN 978-94-011-3050-9 (eBook)
DOI 10.1007/978-94-011-3050-9
1. Photochromic polymers. 2. Photochromic polymers — Industrial
applications.   I. McArdle, C.B.
QD382.P45A87   1991
620.1'92 — dc20   91–18899
                                                   CIP

Typesetting by Syarikat Seng Teik Sdn. Bhd., Malaysia

Preface

Photochromic polymer systems are of two main types: those which are merely solid solutions of photochromes in polymeric matrices and those custom-designed polymers which inherently exhibit photochromism. This book provides a concise review of developments in such systems over the past two decades. The coverage has been limited specifically to applied systems, or areas with potential applications, although over 500 references cite much of the literature on the fundamentals of the subject.

In general, non-biological organic photochromism in organic matrices has been covered. However, the unique properties of polysiloxanes merit special mention in Chapter 4, because of the attributes that such inorganic polymers can provide in certain systems such as liquid crystalline photochromic polymers, where two extremely interesting phenomena are combined. In addition to outlets in polarization-sensitive holographic recording media, such materials exhibit interesting non-linear optical effects suitable for optical switching and rheo-optical phenomena which may find application in mechano-optic transduction. Within this framework examples of all the important photochromic mechanisms are covered by authors from both the industrial and the academic sectors.

Given the photonic nature of the phenomenon under discussion, it is not surprising that many optical applications have been proposed. It is perhaps more surprising, however, that, until recently, no large scale markets had been identified that could commercially exploit photochromic phenomena. Invariably, the better candidates for exploitation involve photochromic polymer systems. Many arguments can be put forward to explain this slow uptake of photochromic 'device' application. These would include poor performance with regard to fatigue, thermal reversibility, light fastness, viable systems integration as well as market strategies which may question the very need for reversion of an initially useful/exploitable photochemical conversion reaction. Chapters 1, 2 and 3 of this book broadly address such issues and illustrate that a choice of materials currently exists, for example, to provide stable photochromic systems which may not thermally revert, as for fulgides and related species, and that high performance reversible spiroxazines in appropriate matrices are available which address specific large volume markets, e.g. ophthalmic lenses, only because the combination of all of the properties unique to such systems conform to market needs. These chapters also include a critical analysis of digital, analog and holographic optical storage applications and describe materials suited to passive component fabrication in guided wave optics and active media for optical signal processing.

Chapters 5 and 6 complement earlier chapters in the book by dealing with areas which require research investment for future applications. The penultimate chapter broadly discusses photoresponse in polymers, from conceptual perspectives through to impressive working examples, and allows the reader to follow experimental strategy clearly throughout. This leads, ultimately, to identification of the somewhat previously neglected triphenylmethane-type photochromism in polymers, as exemplary candidates for the demonstration of large (field-assisted) genuine photomechanical responses which may have future applications, e.g. in novel photomechanical transducers. Chapter 6 provides a definitive overview of redox photochromism in viologen-based solid state systems. From an applications standpoint, such systems already describe impressive photochromism which gives rise to absorption at near infrared wavelengths, a region easily accessible to current laser diode technology. This chapter should stimulate new research in a less common area of photoresponsive polymers, so that, when combined with the other chromogenic phenomena available to such systems, new applications may emerge.

This volume complements existing and recent books on photochromism, which deal with low molecular weight photochromic systems mainly from a more fundamental perspective. It should appeal to industrial and academic researchers in materials science, chemistry, physics and electronics.

The editor is indebted to the contributors, experts in their field, for their superb co-operation throughout this project. Finally, as always the editor warmly acknowledges the encouragement given by his wife.

C.B.McA.

Contributors

Dr J.C. Crano PPG Industries Inc., Chemicals Group Technical Center, 440 College Park Drive, Monroeville, PA 15146, USA

Professor M. Irie Institute of Advanced Material Study, Kyushu University, Kasuga-Koen 6–1, Kasuga, Fukuoka 816, Japan

Professor H. Kamogawa Yamanashi University, Faculty of Engineering, Department of Applied Chemistry, Takeda-4, Kofu, Japan

Professor V. Krongauz The Weizmann Institute of Science, Department of Structural Chemistry, Rehovot 76100, Israel

Dr W.S. Kwak PPG Industries Inc., Chemicals Group Technical Center, 440 College Park Drive, Monroeville, PA 15146, USA

Dr C.B. McArdle Loctite (Ireland) Ltd, Research and Development Laboratories, Chemistry and Materials Science Department, Tallaght, Dublin 24, Republic of Ireland

Dr C.N. Welch PPG Industries Inc., Chemicals Group Technical Center, 440 College Park Drive, Monroeville, PA 15146, USA

Dr J. Whittal University of Wales, School of Chemistry and Applied Chemistry, PO Box 13, Cardiff, CF1 3XF, UK

Contents

1 Optical applications of organic photochromic polymer systems

C.B. McARDLE

1.1 Introduction

Photochromism is a reversible transformation of a chemical system between two states whose absorption spectra are distinguishably different, the change being induced in at least one direction by electromagnetic radiation. This is illustrated schematically in Figure 1.1.

Radiation λ_1, acting on species A, generates species B until a photostationary state is reached. In the dark, the photogenerated B forms may thermally revert to parent A forms. This back reaction can also be driven photochemically by irradiating with the wavelength at the maximum absorption of B, whence a new photostationary state will be achieved. The most common photochromic reactions are of the positive or normal type, in which the initial chemical system comprises a unimolecular A form which absorbs at a shorter wavelength (usually near UV/blue) and a B (or coloured) form that absorbs at wavelengths in the visible spectrum. Bimolecular chemical systems as well as negative or reverse photochromism ($\lambda_{max^A} > \lambda_{max^B}$) are also possible in organic systems. Unless otherwise stated this chapter is presently concerned with unimolecular, positively photochromic systems confined within amorphous or ordered solid or liquid crystalline states.

It is important to note that photochemically produced B forms need not be susceptible to thermal reversion to A in the dark. Whilst the former is quite a common feature, it is those systems which have been specifically molecularly engineered to be thermally stable in the B form, and reverted exclusively by photochemistry (driven by λ_2 irradiation), which have the greatest potential for practical device application.

The simplistic notion given thus far for A $\rightleftarrows$ B interconversion ignores the fact that further undesirable side reactions involving the reactivity of B with active species, e.g. H_2O, O_2, etc., can result in photoinduced decomposition and ultimately fatigue and loss of photochromism in the system. Fatigue and cyclability, i.e. the number of times the interconversion can be made without significant performance loss, as well as the photochemical quantum efficiency for forward and reverse (and decomposition) reactions,

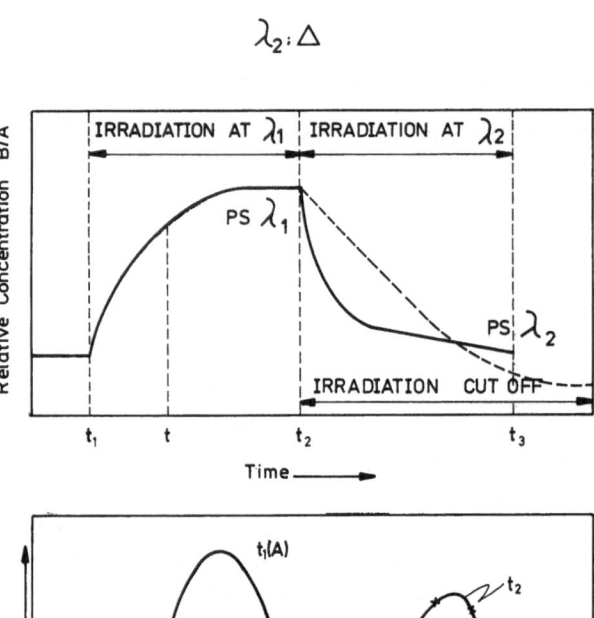

Figure 1.1 Schematic representation of the typical behaviour of a positively photochromic unimolecular reaction. A converts to photostationary state PSλ_1 on irradiation at λ_1 with concomminant spectral change described over time period t_1 to t_2. The intermediate stage t recorded before the photostationary state PSλ_1 is also shown. (Redrawn from Eisenbach, 1979.)

are all important physical parameters which require characterization for device application.

Many organic (and inorganic) systems exhibiting photochromism as well as interesting chromogenic phenomena such as thermochromism (thermally-induced reversible colour change), solvatochromism (solvent-induced change), piezo- or barochromism (pressure-induced change), sorpto-chromism (acid–base-induced change), electrochromism (E-field-induced change) and magnetochromism (H-field-induced change) have been detailed or reported in the literature (Brown, 1971; Lampert and Granqvist, 1990).

With regard to photochromic phenomena in organic molecules, six general categories may be used to classify the mechanisms responsible for the effect in the majority of cases, *viz.* heterolytic/homolytic bond cleavage, *cis–trans* isomerization, valence tautomerism, electron transfer systems, pericyclic reactions and triplet–triplet absorption. Several of these systems will be discussed throughout this text, particularly as inherent parts or components within macromolecular systems.

In this chapter it is assumed that the reader is familiar with the basic photochromic reactions of spiropyrans (heterolytic cleavage), fulgides (tautomerism) and azobenzenes (*cis–trans* isomerization). Extensive fundamental studies on these and other systems are documented in the works of Brown (1971), Japaridze (1979), Dürr (1989) and El'tsov (1990), and spiropyran-like and fulgide–fulgimide systems are further detailed in Chapters 2 and 3 of this book. From an applications standpoint the texts of Dorion and Wiebe (1970), Brown (1971) and Barachevskii *et al.* (1977) are also recommended.

The purpose of the present chapter is to comment on selected aspects of photochrome confinement in polymer matrices from a device viewpoint, as well as critically to review selected aspects of proposed optical applications for photochromic polymer and related systems over the last two decades. Because of the dramatic and reversible changes in the optical properties (absorption, refractive index) of a photochromic material on excitation, it is not surprising that much international research effort has centred on applications aimed at optical recording as well as in light attenuation control, for example in architectural and automotive glazing as well as in eyeglass wear. The latter area is close to commercialization and is described in detail in Chapter 2. Optical recording exploiting photochromic polymer systems has not, however, made an impact in the marketplace, and some of the reasons for this are alluded to in the present chapter, wherein systems considerations and competitive technologies are taken into account. It would appear, however, that the prospects for application in specialized holographic outlets, where hardware constraints are not so severe and photochemical reversibility and high resolution can all be used to full advantage, might become important in the future.

1.2 Potential for erasable optical storage in photochromic polymer systems

From the two available states (nominally colourless and coloured) associated with common photochromic processes, it will be clear that there is a basis for erasable optical storage/imaging systems which exploit this phenomenon.

Photochromic media for this application may be conveniently prepared by incorporation of the photochrome into a polymer binder in the form of a solid solution; alternatively, the polymer itself may be inherently photochromic. The former represents a facile low-cost option, whilst the latter may offer advantages from a stability viewpoint as well as the possibility of systems advantages (see p. 8). In either case, well-known techniques for high-quality film preparation used in the coatings industry should be directly applicable to photochromic media preparation.

In glassy solid solutions absorbances of 2 or more can be realized in films a few microns thick and at low photochrome loadings (approx. 2–3%). Nevertheless, care should be taken to avoid aggregation effects which can adversely influence kinetic behaviour as well as the spectral characteristics of the media. For example, Jones *et al.* (1989) noted that an approximately 2% solution of fulgides in toluene developed an additional absorption band on standing and have tentatively assumed this to be the result of dimerization. In their studies on photoanisotropic effects in solid poly(methyl methacrylate) (PMMA) solutions they have used approximately 0.25% fulgide levels to avoid potential problems.

Krongauz and his co-workers (Goldburt *et al.*, 1984; Goldburt and Krongauz, 1986) have reported new crystallization phenomena in atactic vinyl polymers bearing spiropyran groups in films cast by evaporation from methyltetrahydrofuran solutions. The so-called 'zipper crystallization' results from the mutual stimulation of solvatochromatically induced spiro → merocyanine interconversion and crystallization. There appears to be no reason to assume that this phenomenon does not occur in simple solid solutions of spiropyrans in polymer matrices (Ekhardt *et al.*, 1987). The presence of such aggregates formed by antiparallel dipole–dipole interaction in merocyanines can occur at relatively low loadings (approx. 5%) and has a pronounced effect on decolourization kinetics (retardation) and causes hypsochromic shifts in the absorption maximum of coloured forms. Merocyanine association phenomena can lead to very stable coloured forms in crystalline (Goldburt *et al.*, 1984) and liquid crystalline polymeric systems (Yitzchaik *et al.*, 1990) (see also Chapter 5).

Aggregation effects are also known in azobenzene systems, so that while it is true to say that solid solutions of a variety of photochromes in polymer binders can be conveniently prepared care should nevertheless be taken to avoid association phenomena and their consequences on kinetics and wavelength tuning-in devices.

Although the concept of memory devices based on photochromism was proposed more than 30 years ago (Hirshberg, 1956), photochromic recording media have only reached the marketplace in an intermediary capacity via an early micrographics system [PCM I (NCR Co.), see Carlson *et al.*, 1962] and in small market niches for specialized films used in graphics work, etc. (e.g. Kodak photochromic film SO-548, Cyanamid films 43–540 (A),

51–142). There have been applications exploiting a single cycle from a photochromic reaction in instant imaging (Böttcher and Epperlein, 1983) but none has been realized in, for example, erasable digital storage despite the intense international research in photochrome development and the encouraging characteristics of write once–read–many (WORM) as well as some erasable alternative dye-in-polymer systems (Alexandru et al., 1984; Gupta and Strome, 1985; Hartman and Lind, 1987). The principal reason for this situation would appear to relate to systems constraints since modern photochromes that exhibit good intrinsic fatigue characteristics and possess thermally stable coloured forms have been developed (see Chapter 3).

It is pertinent to consider briefly some of the very fundamental requirements, or highly desirable characteristics, for commercially viable, digital, bit-oriented optical memories in order to assess how likely photochromic polymer systems may conform. Desirable features include: high-integrity media (preferably with threshold characteristics) capable of facile mass production and of being handled under normal ambient conditions (light, O_2/H_2O, room temperature, etc.) without loss of performance (or data); simple optical systems, preferably using a single, low-power (approx. 5–10 mW), near-infrared (approx. 780–850 nm), directly modulated laser diode capable of providing all write, non-destructive read as well as tracking and focus error correction functions; high data rates (10 Mbits/s); high sensitivity ($\leq nJ/\mu m^2$); high bit resolution (approx. 1 μm^2/bit); low bit error rates ($\leq 10^{-7}$); efficient implementation of fast data verification techniques such as direct read after or during write (DRAW or DRDW respectively); long-term bit stability (~ 10 years), etc. (see Bouwhuis et al., 1985). Erasable media further require high cyclability (~ 10^6 cycles) and preferably single-laser operation and similar kinetics for write and erase functions.

Notwithstanding the non-exhaustive list of requirements set out above, there is some question as to the relative virtues of erasable optical storage. The impetus to consider optical-based technology in the first instance was founded largely on the huge storage capabilities available to such systems resulting from focused coherent monochromatic light. Thus, WORM optical discs offer massive redundancy in memory which somewhat negates the need for erasability whilst at the same time offering the decided advantage of archivability.

At present it is impossible for photochromic media to fulfil many of the basic requirements for erasable digital recording in a commercially viable system. Even if ideal absorption characteristics and convenient wavelengths are assumed, many fundamental problems still remain, particularly from a kinetics standpoint. The kinetics of photochromic reactions (primarily thermal decolourization) in glassy polymer matrices has been extensively studied and comprehensively reviewed (Smets, 1983). Various studies have revealed that above the glass transition temperature (T_g), thermal back reactions obey a single first-order rate, whilst below T_g the behaviour is usually more complex. It is generally accepted that this is the result of an inhomogeneous distribution of free volume within the matrices [Smets (1983) and references therein], although other possible mechanisms have been proposed (Goldburt et al., 1986). The findings of Eisenbach (1979, 1980) pertaining to the dependence of photochromic processes in bulk polymers on free volume and their control by molecular motions within the polymer matrix itself has led to the conclusion that it is unlikely that the phenomenon has a good future in (rapid, reversible) photoimaging processes since reaction rate is determined by the mobility of the matrix surrounding the chromophore. However, in alternative types of photochromic imaging systems this may not be so serious a problem. More recently, several detailed studies on photochrome kinetics and on probing the size and distribution of free volume in glassy polymers and the effects of annealing photochromic polymer systems have been published (Lamarre and Sung, 1983; Sung et al., 1984; Richert and Baessler, 1985; Victor and Torkelson, 1987a,b, 1988; Richert, 1988; Yu and Sung, 1988; Tsai et al., 1990).

It is somewhat surprising that, in spite of the wealth of kinetics studies on photochromics in polymer matrices and the apparent interest in the exploitation of such materials in erasable optical storage, relatively few studies on photoinduced decolourization (even in systems with thermally stable coloured forms), especially in experimental arrangements which might simulate a practical optical system design, have been undertaken. Horrie et al. (1985) have, however, studied pulsed laser (337 nm) induced colouration with (continuous) visible light (filtered, 560 nm) decolouration for a spiropyran solid solution in a polycarbonate matrix. Such a study reveals much information for a systems engineer, e.g. write times in the nanosecond time regime (2.5 mJ, 10-ns pulse) are apparent, as is fluorescence emission occurring during the write pulse, which could in principle be useful as a DRDW verification system technique. However, the erase times are disappointing, occurring on the millisecond time scale, even if it is assumed that the majority of photoinduced decolourization occurs at elevated temperatures because of laser heating. It would not be anticipated that pulsed rather than continuous irradiation would improve this situation dramatically. Furthermore, it is already known that the quantum yields for photodecolourization (at erase wavelengths) for spiropyrans are generally orders of magnitude lower than for photocolouration (Bertleson, 1971; Wu and Yunfei, 1988).

Tomlinson (1984) has analysed the potential of model-generalized photochromic media for bit-oriented memory and has highlighted a significant number of deficiencies, most notably relating to lack of threshold for the phenomenon, erase power requirements, destructive read-out and the system requirements for tracking and focus control, the latter dictating the use of an additional laser at a medium-insensitive wavelength.

Notwithstanding the difficulties in this area, some of which have been highlighted above, the combined developments in photochrome design and electro-optics technologies may change this perspective in the future. A detailed description of a promising class of photochromes developed by Heller and co-workers is given in Chapter 3. Together with high intrinsic fatigue characteristics, fulgide derivatives may be made thermally stable so that interconversion between the two forms is via purely photonic processes. This class of molecules exploits a tautomerism mechanism in its photo-chromism and has a minimum of steric requirements. Irie (1987) and Irie and Mohri (1988) have developed related photochromes such as 2,3-bis-(2,3,5-trimethyl-3-thienyl)-maleic anhydride, the photochromic reaction and derivatives of which are shown in structures **1** and **2** below.

Molecules of this type do not exhibit thermochromic reactions at high temperature (300°C), unlike certain fulgides which become thermochromic above 120°C. However, the quantum yields for the thienyl-type materials' reversible reactions are low. Irie and Mohri (1988a) have shown that the activated forms of dithienylethene derivatives have absorption tails extending to 700 nm and can be bleached with HeNe laser light at 633 nm. Development work continues to provide increased sensitivity in these systems at common laser diode wavelengths. Vinyl functionalized versions (**2**) have also been prepared (Irie, 1988) to provide photochromic homo- and copolymers with spectral characteristics similar to the present molecule. In copolymer films, photoinduced tautomeric ring closure is more efficient at elevated temperature because of increased mobility in the polymeric system, and it has been suggested that this temperature dependence might be exploited in a non-destructive read-out capability.

Very recently Shen *et al.* (1990) have disclosed an interesting class of thermally irreversible photoresponsive copolymers containing side-chain *trans*-cinnamamide chromophores. Progress towards improved near-infrared sensitivity has also been made in systems employing heterolytic cleavage mechanisms, e.g. spirothiopyrans have been developed with coloured-form absorbances of 0.42 at 720 nm and 0.13 at 780 nm, although these species are not thermally stable in their coloured forms (Arakawa *et al.*, 1984). Stabilization of photomerocyanines based on such thio derivatives, even by judicious choice of a polymer matrix, physical treatments to the matrix and photochrome molecular size criteria (see Smets, 1983), would only lead to a medium with mediocre infrared sensitivity but impractical erase kinetics. Thus the design strategy behind shifting the coloured form absorbances into the infrared for thermally revertible photochromes remains questionable. The viologen systems developed by Kamogawa and co-workers and detailed in chapter 6, section 6.8, appear to be the most promising systems currently available with regard to efficient coupling of coloured-form absorbance to near-infrared radiation.

The importance of thermal stability in photogenerated coloured forms for use in practical imaging systems cannot be overemphasized, particularly when the local heating effects of tightly focused laser beams are taken into consideration (Tomlinson, 1984). Even low-power lasers (approx. 5 mW) produce sufficiently high energy densities when focused into small areas (approx. μm^2) to cause temperature jumps of hundreds of degrees for short periods of time followed by rapid quenching. Indeed it is this very principle that ablative, phase change, magneto-optic, liquid crystalline, etc. media exploit in their operation (for example, for the organic liquid crystalline case see Armitage, 1981). The effects of such temperature jumps on photochromic polymer systems require quantification.

The effect of UV radiation on a proposed photochromic polymer matrix for optical recording should also be taken into consideration. In spite of molecular design to limit intrinsic fatigue in photochromes, irreversibility and decomposition effects resulting from reaction with oxygen, water vapour impurities or additives and the polymer matrix itself under irradiation can lead to degradation. In certain cases such reactions have been used to advantage for irreversible imaging. For example, photolysis products from the polymer matrix or additives therein may be used to fix photomerocyanines (Arnold, 1977; Delzenne, 1979; see also Ohno 1975). Likewise the oxygen-promoted phenanthrene formation in the side chains of photochromic polymers bearing stilbene pendant groups (Tsai *et al.*, 1990) constitutes an irreversible system. Nevertheless, these effects are in general undesirable.

In the spiropyran case, it has been found that copolymerization of vinyl functionalized photochromes with methyl methacrylate reduces the quantum yield for photodecomposition by a third compared with simple solid solutions of spiropyrans in a polymer matrix. This is thought to be because of inhibition of the relative rotation of parts of the spiropyran molecule (Kardash *et al.*, 1974). Smets (1975) has also studied the photodecomposition of spiropyran-based polymers and has noted the effects to be considerably less pronounced in the absence of oxygen. Wilson (1984) has clearly highlighted the extent of the problem of oxygen interaction with thermally stable fulgides in solid polymer solutions and has established a correlation between degradation rate and oxygen permeability of the matrix, as shown in Figure 1.2. It was noted that an order of magnitude improvement in the number of colouring–bleaching cycles was seen when fulgides were incorporated in a low-permeable matrix and sandwiched between glass plates compared with the same material coated on a cellulose acetate base. The presence of antioxidants, UV stabilizers, etc., can help the situation, for example organonickel and hindered amine light stabilizers have been used to effect in conjunction with spiroxazines in polymer matrices (Chu, 1985, 1986, 1988; Sugiyama and Sakai, 1987; Furuta *et al.*, 1988). It is interesting to note that the latter types of photochrome, although closely structurally related to spiropyrans, have quantum yields for photodecomposition 10^3 times smaller (see Chapter 2).

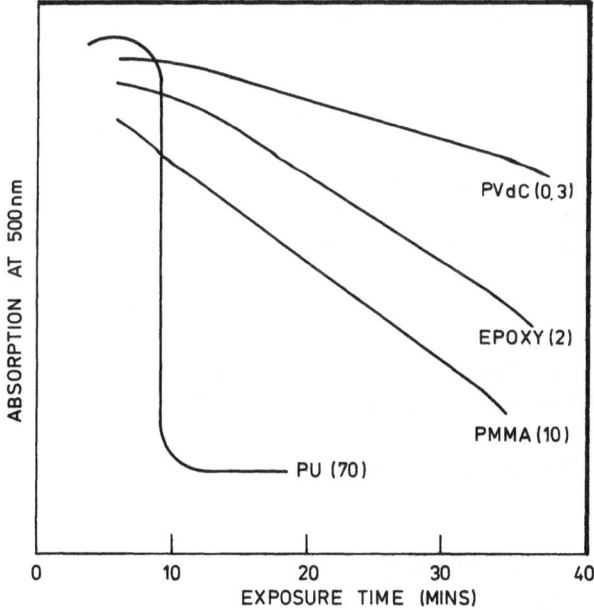

Figure 1.2 Effect of oxygen permeability on photoinduced fatigue in polymer–fulgide systems irradiated at 366 nm and measured as change in absorbances at 500 nm. Figures in parentheses are in $cm^3/cm/cm^2/s^1$ $(cmHg)^{-1}$. Pu, polyurethane; PMMA, poly(methyl methacrylate); PVdC, poly(vinylidene dichloride). (Redrawn from Wilson, 1984.)

Irradiation can also cause defects in the matrix resulting from excess energy absorbed by the photochrome, which when transmitted to the medium can cause conformational changes even in the glassy state of the matrix (Kryszewski and Nadolski, 1977). Overexposure, like chemical reaction, has also been proposed as means of achieving a form of irreversible optical imaging (Hawkins and Bowyer, 1988).

Some general effects of high-energy radiation on polymers have been reviewed by Reichmanis and O'Donnell (1989).

Returning to systems aspects, recent advances in electro-optics technology are noteworthy. Laser diodes have been designed for more conventional optical storage media which have improvements in operating wavelength, reliability and power output as well as multiple lasers on a single chip designed specifically for easily alignable optical heads with write–read–erase capabilities (Hattori et al., 1987; Yamaguchi, 1987; Ueda, 1987; Urita, 1987a, b). Such sources together with developments in efficient optical frequency doublers could afford a potentially useful system capable of addressing (write/erase) the appropriate absorption bands in a purely photochemically reversible photochromic medium. However, many problems remain to be overcome with the latter technology (Engler, 1990).

Preheat-aided laser recording techniques, such as those developed for WORM systems (Goldberg, 1979), could also possibly be used to advantage in a system with non-destructive read-out in, for example, media like the photochromic copolymers of the dithienylene type described by Irie (1988), which require thermal assistance to facilitate photochemical reversion.

Although there appears to be progress in photochrome design with regard to fatigue, thermally stable excited forms and near-infrared absorption characteristics, as well as progress in optics technology, it is clear at present that erasable digital storage for photochromic media is a long way off and may, in fact, never be a practical proposition, particularly when competitive media which are already well developed are taken into consideration (Bouwhuis et al., 1985; Chen et al., 1985; Shieh and Kryder, 1987).

Some of the difficulties associated with the use of photochromics in rapid, reversible, digital optical stores have been described above. If analogue optical recording is now briefly considered, for example in micrographics applications, it becomes apparent that difficulties also exist, but for different reasons, in the full exploitation of the properties available to photochromic polymer films.

Modern micrographics enjoys a healthy market (Coopers and Lybrand, 1987) and computer output to microfilm (COM) and computer-aided retrieval (CAR) techniques are state of the art within this technology. COM systems (Hauser, 1983) take advantage of computer-controllable high-speed laser scanning techniques to address a variety of recording media which may be instantly imaged, e.g. the thermo-optic laser direct-recording film (LDF) film of Fuji (Miyauchi *et al.*, 1983) or dry developable silver technology (Morgan, 1979). Major markets exist for such recording media and for unconventional laser recording materials in general (Gillespie and Lee, 1979; Plumadore and Seiteri, 1979). However, erasability in the vast majority of such systems is a distinct disadvantage. This is a consequence of the intended use of such media in permanent human-readable record-keeping, typically of legally sensitive documents such as audits, medical and personnel records, confidential reports, insurance files, etc. Clearly the ability to erase and rewrite to such films is unacceptable. On the other hand, updatability may be considered an asset. This feature allows information to be added on to a section of film, e.g. a microfiche, which already has some information recorded onto it. For example, a partially recorded silver halide fiche, once developed, cannot be updated because development causes loss of light sensitivity to the entire film, including unimaged areas. Updatable microforms fill a market niche. The most successful type, transparent electrophotographic (TEP) film, exploits organic photoconductive phenomena but still requires wet chemical development (Shaffert, 1975; Palm *et al.*, 1979). A photochromic film which exploits fixation techniques to stabilize permanently a coloured form could offer some advantages as a dry-imaging, laser-addressable, updatable medium.

The photosensitivity of the master form of the proposed film might, however, be more problematic than in the TEP case (the latter tends to be infrared-sensitive). Various potentially useful fixation techniques have been described for photochromics in previous literature (Dorion and Weibe, 1970; Brown, 1971; Jacobson and Jacobson, 1976; Delzenne, 1979; Heller, 1982; Goodman and Parsons, 1983; Reid and Wafers, 1983; Tamura and Seto, 1980; Yoshida *et al.*, 1987).

Holographic optical storage/application is another avenue open to photochromic media. Markets for conventional holographic products are beginning to grow (Clarke, 1987; Ritzau, 1987) and applications in the industrial sector include: head-up display, holographic optical elements, real-time data storage media and elements for optical signal processing. Erasability in holographic systems can be used to advantage in these three application areas. For early literature on materials parameters relevant to recyclable holography, the reader is referred to the reviews by Bordogna *et al.* (1972), Lo and Honebrink (1972), Kvasnikov *et al.* (1973), Kurtz and Owen (1975) and Tomlinson and Chandross (1979).

Some unusual holographic applications for photochromic polymer films, photochromes in polymer fluids or tagged biopolymers in solution have been described for mapping microwave and acoustic fields (Iizuka, 1971, 1972), flow measurement (Fowlis, 1979; D'Arco *et al.*, 1982) and in the study of biological systems (Mlles *et al.*, 1983).

Selected applications in the areas of optical signal processing and erasable optical elements using photochromic holographic media as well as photochromic imaging in ordered systems are detailed in the following section.

1.3 Photochromic imaging in ordered systems or in systems exploiting ordering effects

Imaging concepts based on the consequences of photoinduced transformations have been described in a variety of ordered systems, many of which are based on photochromic species possessing long hydrocarbon chains. For example, a molecularly dispersed state may be converted to a micellar state, or the shape of given micelle may be photomodified in surface-active spiropyrans and azobenzenes, and these effects have been described as being of relevance to optical memory systems and amplified image recording (Tazuke *et al.*, 1987a,b). Similarly derived spiropyrans as well as photochromic polymer systems have been the subject of extensive investigation in Langmuir monolayers (Polymeropoulos and Mobius, 1979; Gruler *et al.*, 1980; Morin *et al.*, 1980; Holden *et al.*, 1984; Vilanove *et al.*, 1983; McArdle and Blair 1984; Blair and McArdle 1984a,b) and in Langmuir–Blodgett films employing anthrocyanidins (Möbius *et al.*, 1969), spiropyrans (Polymeropoulos and Mobius, 1979; McArdle *et al.*, 1983; Ando *et al.*, 1985, 1988), stilbenes and thioindigos (Whitten, 1979), azobenzenes (Fu Kuda and Nakahara, 1978; Tachibana *et al.*, 1989) and viologens (Nagamura *et al.*, 1990), and patent applications based on this type of technology applied to laser-addressed optical recording devices have been issued (Nakagiri *et al.*, 1986a–d; Miyazaki and Ando, 1986). This approach to device application is not, in the author's opinion, viable from performance, cost or fabrication standpoints. A much more flexible approach involves the use of monomeric and polymeric liquid crystalline (LC) media in conjunction with photochromic species. Interestingly, the advent of a photochromic liquid crystal was predicted by Bertelson in 1971. In the monomeric LC case, photochromes have been included as guest molecules in the bulk mesomorphic host, as responsive species in LC alignment layers, or as an inherent part of the LC molecule itself. In the polymeric case, solid solutions of photochromes in liquid crystalline polymers (LCPs) or as copendant moieties with mesogenic species have also been studied, as will be discussed in detail later. A brief discussion on the monomeric LC host types will be given first to demonstrate the functionality of these systems.

Ogura *et al.* (1982) have demonstrated a novel LC-based UV imaging device. In this case inclusion of an azobenzene in a smectic LC host enables a photoinduced smectic → nematic phase transition to be realized. The device was configured with homogeneous alignment layers and an initially scattering smectic focal conic optical texture. On UV irradiation the more easily aligned nematic phase is generated and spontaneously aligns in a planar disposition in the locally illuminated zones. Exposure to 440 nm light back-converts the aligned nematic regions to storage-stable planarly aligned smectic regions which now produce contrast against the scattering background. Because the magnitude of photoregulation of the phase transition is relatively small, the device requires thermal biasing. The phenomenon results from the co-operative disruption/restoration of bulk ordering in the LC host by the photoisomerizing guest molecules, an effect which has also been exploited in polymers, as shall be discussed later.

Ichimura *et al.* (1988) have described an elegant photocontrollable switching/ memory effect in a nematic liquid crystal device by using a photochromic azobenzene surfactant as an alignment layer capable of switching between homeogeneous and homeotropic configurations which in turn controls the bulk orientation of the overlying LC. These authors have estimated that two isomerizing azobenzene units can control the orientation of about 15 000 LC molecules, which because of their highly birefringent characteristics can produce large changes in optical transmission through crossed polarizers. A schematic representation of how these photochromic 'command surfaces' function is given in Figure 1.3. Conceivably such effects could be employed as inexpensive, large-area, optically addressed, spatial light modulators which might compete against the more established and more complex photoconductor–LC type. It would be of interest to examine their performance in this regard.

The command surface concept has been extended to include photochromic azobenzene polymers as LC alignment layers (Ichimura *et al.*, 1989).

Read-out of reoriented LCs by photoinduced control of alignment need not exclusively employ optical methods. If, for example, LCs with large

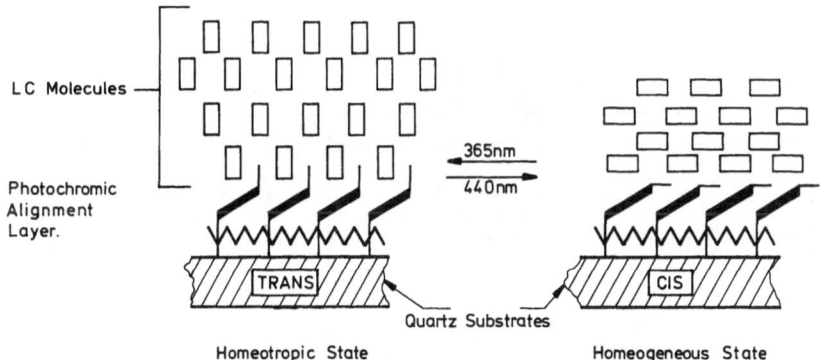

Figure 1.3 Schematic representation of the function of a photochromic alignment layer (polymer or surfactant) for commanding the orientation of bulk liquid crystals. (Redrawn from Ichimura *et al.*, 1988.)

dielectric anisotropies are used, a non-destructive capacitance technique could be employed to good effect in Ichimura's system allowing for an optoelectronic transduction system. The modulation of dielectric constant in an LC device induced by photocontrol of the nematic to isotropic phase transition has already been demonstrated by Kurihara *et al.* (1988), where small changes in capacitance have been noted in azobenzene-doped systems. In this case response times were slow, although it was suggested that this situation could be improved in ferroelectric LCs. However, in the author's opinion this is unlikely because of steric constraints in these more ordered phases and the range of currently available transitions in such LCs.

A further variation on the combined photochromic/mesomorphic systems has been described recently by Kimura *et al.* (1989a, b), who have exploited a photoinduced ionic conductivity jump in polymer films containing an azobenzene species which was inherently liquid crystalline in an electrostatic imaging technique. Thus photoinduced transformation of the azobenzene species from a nematic to an isotropic state within an ionically doped polymer binder gives rise to an enhancement of ionic conductivity by two orders of magnitude where locally irradiated. This charge compensates surface charge loaded by corona charging, thereby allowing subsequent electrostatic imaging by conventional xerographic methods. The UV-generated image on the photochromic polymer composite can subsequently be erased by visible light and the process repeated. Figure 1.4 describes the concept of this novel imaging method.

Some of the aforementioned as well as additional effects have been observed when the matrix confining the photochrome is a polymeric liquid crystal, and some very useful storage effects have emerged recently. By analogy with the photoinduced phase transition effects already noted in low-

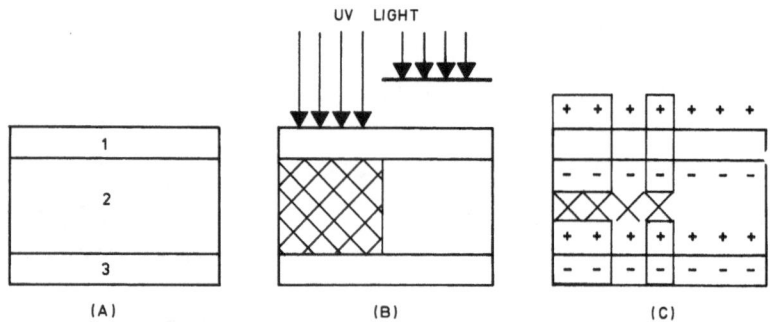

Figure 1.4 Concept of image storage in a photochromic/LC composite film. (a) Device structure: (1) insulating film; (2) ion-conducting photosensitive polymer/LC matrix; (3) electrode. (b) Exposed region (UV light) becomes highly conducting. (c) Subsequent corona charge load compensated by charge in regions with enhanced conductivity — residual charge in dark regions gives latent electrostatic image. (Redrawn from Kimura *et al.*, 1989a.)

molar-mass LCs, similar effects have been reported in mesogenic copolymers containing azobenzene-type dopants or pendant side chains. It would appear that thermal biasing of the sample is required during recording of information. Read-out in such systems is accomplished by observing the disappearance of birefringence through crossed polarizers following light-induced change from the nematic to the isotropic state. Stored information can be frozen in by cooling the sample below the polymer glass transition. A resolution of 2–4 μm has been reported in such systems (Ikeda *et al.*, 1988, 1990a, b).

Photoinduced dichromism and birefringence in a variety of dye–matrix systems have been extensively studied relatively recently, and such materials constitute an important class of polarization-sensitive media. Examples of such media include the triplet photochromism of xanthene dyes in an orthoboric acid matrix (Shankoff, 1969) and photochromic fulgide and azobenzene derivatives in conventional and in liquid crystalline polymers (Todorov *et al.*, 1986; Eich and Wendorff, 1987; Eich *et al.*, 1987; Jones *et al.*, 1989; Kozak and Williams, 1989; McArdle, 1989).

Photodichroism was demonstrated by Weigert in 1919 in silver chloride emulsions which became strongly dichroic after pre-exposure to natural light and further illumination with red polarized light (Weigert, 1919). Illumination by natural light creates a so-called silver–silver chloride (Ag–AgCl) emulsion which exhibits a wide red absorption band due to the existence of photolytic silver. The additional illumination with red polarized light induces photoadaptation, i.e. the selective bleaching of part of this band and an optical anisotropy. Photodichroic glasses offer serious competition to conventional photochromic systems in device application, and some of the attributes of the former type of systems should be mentioned before describing photodichroic dye-in-polymer systems.

Photodichroic glasses may be used as digital, direct read after write (DRAW), laser recording media which permit instant optical recording without processing of any kind. They are also suitable for high-resolution holographic media suitable for storage and coherent optical computing elements (Caimi *et al.*, 1976).

In photodichroic silicate glasses, UV irradiation generates strong absorbance due to colour centres (Ag–AgCl). Exposure of the initially optically isotropic sample to polarized light in the wavelength region 0.5–0.7 μm induces dichroism and birefringence over a broad wavelength regime, 0.4–1.0 μm. In these systems, the optic axis of the induced anisotropy coincides with the polarization direction of the incident beam. This direction may be arbitarily chosen since the alignment of the transition dipoles is not along a crystalline axis. Writing is performed with linearly polarized light at the bleaching wavelength and reading is performed at a wavelength outside the absorption band and exploits induced birefringence effects. Erasure is effected by re-exposing a bit with the write beam whose polarization direction has been rotated 45° with respect to the initial polarization exposure direction. Such systems offer many advantages over photochromic recording media. The initial photocoloured state is not darkened under conventional lighting and shows no fading at elevated temperatures. The system shows no fatigue and may be non-destructively read out. It exhibits a threshold and there is a large dynamic range for applications which require grey scale. Erasability, updatability and archivability are all possible *in the same system* by use of judicious exposure techniques. It has been demonstrated that through proper choice of glass composition, the photosensitive layer may be created beneath a non-photosensitive protective layer with no additional operation in the fabrication process.

Manufacture in general is simple and lends itself to mass production techniques with no restrictions on physical dimensions of devices. The required writing energy density in photodichroic silicate glasses at MHz and higher data rates is approximately mJ/cm^2 (c.-k.Wu, 1979; Borrelli and Young, 1979).

Polarization-sensitive media allow recording of the polarization of light and its reconstruction by holographic techniques, and a new branch of holography — polarization holography — has emerged as a result of this phenomenon, which has been reviewed by Todorov *et al.* (1986) in organic matrices. New polarization-sensitive materials based on rigid solutions of azo dyes in polymer matrices constitute high-efficiency, sensitive polarization holographic media which allow multiple re-use and thus address many of the shortcomings of certain inorganic photodichroic systems. Todorov *et al.* (1986) have studied two particularly suitable photoanisotropic materials based on methyl red (MR; R = CH$_3$, R$_1$ = H, R$_2$ = CO$_2$H, **3**) and methyl orange (MO; R = CH$_3$, R$_1$ = NaO$_3$S, R$_2$ = H, **4**) in poly(vinyl acetate) and polyacrylates, and in poly(vinyl alcohol) (PVOH) respectively.

Structure	R	R$_1$	R$_2$
3 MR	CH$_3$	H	CO$_2$H
4 MO	CH$_3$	NaO$_3$S	H

On absorbing 488 nm argon laser light, such azo dyes undergo *trans–cis* isomerization. If the light is linearly polarized the molecules tend to order themselves in such a way that the direction of their optical transition is

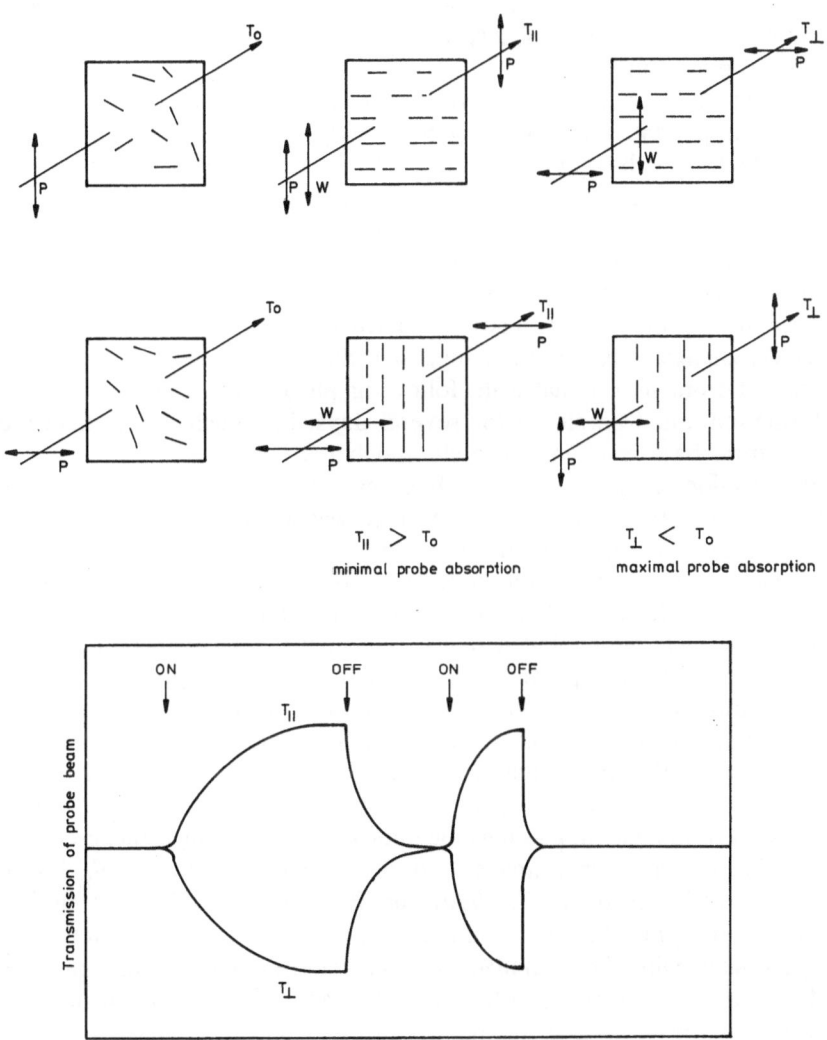

Figure 1.5 Schematic representation of optical transmission change of probing light in a photoanisotropic medium. P, polarization of probe beam; W, polarization of write beam (may be same λ). Write beam comes on and off as indicated. Kinetics of the photoinduced dichroism are faster in subsequent cycles. (Amended and redrawn from Todorov *et al.*, 1986.)

perpendicular to the polarization direction of the light (opposite to the silicate glass case). Thus, the transmission of probing light, polarized along the polarization direction of the writing beam, increases (i.e. absorption decreases); for probing light polarized perpendicularly to the writing beam, transmission decreases, and thus dichroism is induced, as illustrated schematically in Figure 1.5. The kinetics of dichroic and related effects speed up in subsequent cycles, probably as a result of preorientation effects in the photochromic material.

Photoinduced birefringence (Δn) accompanies this dichroism. This is particularly important from a holographic standpoint, at wavelengths outside the photochrome absorption band. It can be measured by following leakage of light (say at 633 nm) through crossed polarizers following a write event (say at 488 nm) which induces photoisomerization and the subsequent transient dye ordering. It was found in MO/PVOH that photoinduced birefringence (Δn) depends on write intensity, dye concentration and the pretreatment of the polymer matrix (Todorov et al., 1984). The unexpectedly large Δn changes (approx. 10^{-3}) suggested that the polymer matrix itself is perturbed from its original state following photoisomerization and phase information could be stored for several days, i.e. much longer than the $cis \rightarrow trans$ back reaction of azo chromophores.

Polarization holographic recording can be accomplished in such photochromic organic media by two plane waves with orthogonal polarizations at the write wavelength. The induced optical anisotropy is spatially modulated in accordance with the polarization modulation of the recording light field. Diffraction efficiency is highest (approx. 35%) when birefringence effects are exploited rather than making use of dichroic effects (diffraction efficiency approx. 1%). The high diffraction efficiency (approx. 35%) allowed the Russian workers to record two separate holograms at the same spatial frequency in the same volume of a photoanisotropic medium. Depending on the polarization of the read-out light, the two images can be reconstructed individually or simultaneously and their common part can be subtracted from their sum — thus the optical logic operations AND, EXCLUSIVE OR can be performed upon images with these media (Todorov et al., 1985).

Eich and Wendorff (1987, 1990) and Eich et al. (1987) have studied photonic recording in photochromic liquid crystalline polyesters and polyacrylates with the formulae and characteristics shown in structures **5–8**. In these systems recording is achieved by the realignment of a uniaxially birefringent LCP monodomain film by optically induced trans–cis isomerization. The monodomain is achieved either by field alignment or more conveniently by annealing treatments on homogeneous alignment layers. Test patterns and gratings demonstrating high resolution (0.3 µm) have been recorded as phase objects in such media. Optically induced birefringence as high as 0.01 and diffraction efficiencies of 50% in thick

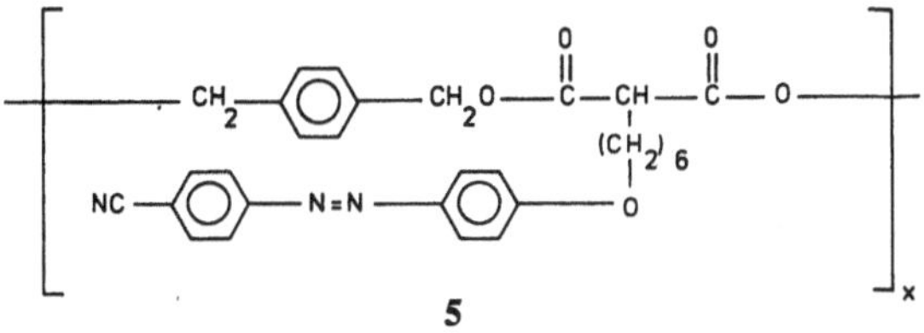

$$g\ 43S_A\ 94n\ 104i$$

5

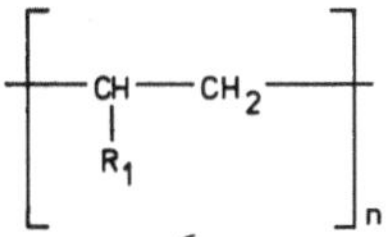

6 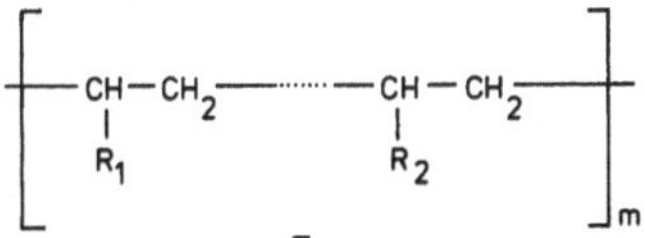   **7**

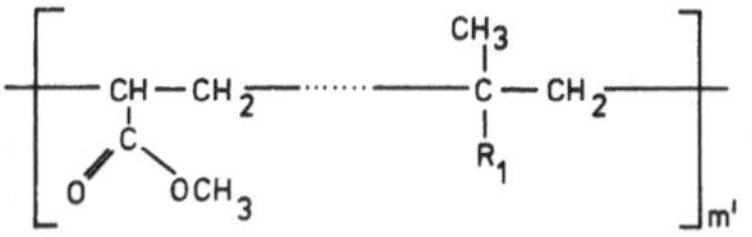

8

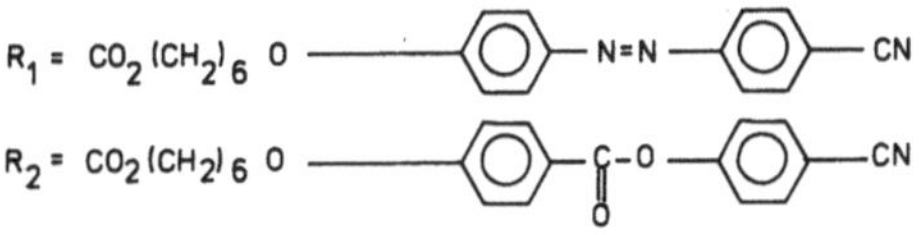

$R_1 = CO_2(CH_2)_6\ O$ — N=N — CN

$R_2 = CO_2(CH_2)_6\ O$ — C-O — CN

STRUCTURE	X (R₁)	X (R₂)	X (CO.OCH₃)	STATES
6	1.00	—	—	~g 30n 110°C
7	0.01	0.99	—	
8	0.25	—	0.75	AMORPHOUS

phase gratings have been measured (Eich and Wendorff, 1987). The intensity required to store information in these photoanisotropic media was about 1 mW/cm².

In addition to the optical processing capabilities referred to above for photoanisotropic media using conventional base polymers, holographic optical elements can also be realized. Thus a radial-symmetric variation of refractive index has been recorded in the LC polyacrylates (see **6–8**) corresponding to Fresnel zone plates, by interfering spherical and planar wavefronts in the medium. These zone plates act as lenses. The functionality of the LCP media enables any particular lens action to be erased and new lenses reformed with different refracting powers, etc. (Eich and Wendorff, 1987).

More recent studies on these materials are aimed at demonstrating that irradiation of a photochromic LCP actually reorients the optic axis of the LCP by 90° from the uniaxial state by illumination with light polarized parallel to the nematic director (Anderle *et al.*, 1989).

Reversible holographic storage has also been demonstrated in azobenzene containing cholesteric cyclic siloxanes (**9, 10**) with T_gs of approximately 50°C and clearing temperatures of around 200°C (Ortler *et al.*, 1989). As with the former systems, these offer the advantage of facile monodomain formation without the need to resort to external aligning fields (E or H). High diffraction efficiencies (up to 40%) were noted in these systems, as were memory retention effects on time scales longer than the thermal back reactions or *cis–trans* isomerization noted by Eich and Wendorff (1987), Eich *et al.* (1987) and Todorov *et al.* (1986).

Williams and co-workers (Jones *et al.*, 1989; Kozak and Williams, 1989) have recently reported angular-dependent photoselection and optical anisotropy for photochromic fulgides in glassy acrylic polymers. Interest-

Structure **9** $R_1 = R_2 = R_3$ as   2 : 1.6 : 0.4 $T_g = 50$; $T_c = 192°$ C.
Structure **10** $R_1 : R_2 : R_3$ as   2 : 1.8 : 0.2 $T_g = 49$; $T_c = 206°$ C.

T_c is transition to cholesteric LC phase.

ingly the fulgide E-α, 2,5-dimethyl-3-furyl-ethylidene-(isopropylidene)-succinic anhydride, was chosen as a photochrome to avoid thermal reversion or irreversible side reactions during their studies. A frozen distribution of the photochromic molecules could be produced by anisotropic photobleaching using polarized laser light.

The authors recognized grey-scale capabilities (cf. silicate glasses) due to the angular dependence of optical properties (dichroism, Δn) in their system and further noted that the principle of angular-dependent photoselection, as it relates to optical information systems, should be applicable to any molecule which can undergo photochemical transformation, reversible or otherwise (Jones et al., 1989; Kozak and Williams, 1989).

The combination of spiropyran photochromism with the properties of mesomorphic polymers has also been addressed largely by the works of Krongauz et al. which are reviewed separately in Chapter 5.

1.4 Applications in optical signal processing and in integrated optics

Bennion et al. (1983) have elegantly described the diversity of ways in which thermally stable photochromic fulgide–polymer systems may be used in passive guided wave applications in integrated optics. The photochromic polymer may be deposited as a thin film on low- or graded-refractive-index substrates by conventional solution coating procedures, alternatively the polymer film may be melted between two substrates. Two-dimensional (UV) laser scanning to the photochromic film enables the generation of complex high-refractive-index waveguide patterns in a one-stage delineation process. Such wavelengths are capable of confining light of wavelengths well outside the photochromic absorption band such as the 1.3–1.6 μm wavelengths used in high-bandwidth telecommunication systems employing monomode fibreoptics. Efficient coupling to the 8- to 10-μm-diameter monomode fibres may be achieved by use of graded-index substrates which carry the bulk of the propagating light in a fashion dictated by the overlying 2–3 μm photochromic waveguide. A significant advantage of these systems resides in the fact that an array of fibres may be linked to the 'latent optical circuit', i.e. prior to waveguide generation. Subsequently, accurate beam-steering techniques may be used to align precisely the UV-generated waveguides to the prepositioned fibres rather than having to align fibres to predefined waveguides, which is much more difficult and is the norm in alternative systems.

Figure 1.6 is a schematic representation of some of the possibilities available in passive optical circuits employing photochromic waveguides. Situation A pertains to a colourless photochromic polymer-coated substrate with a prepositioned monomode fibre array. Situation B represents the sub-

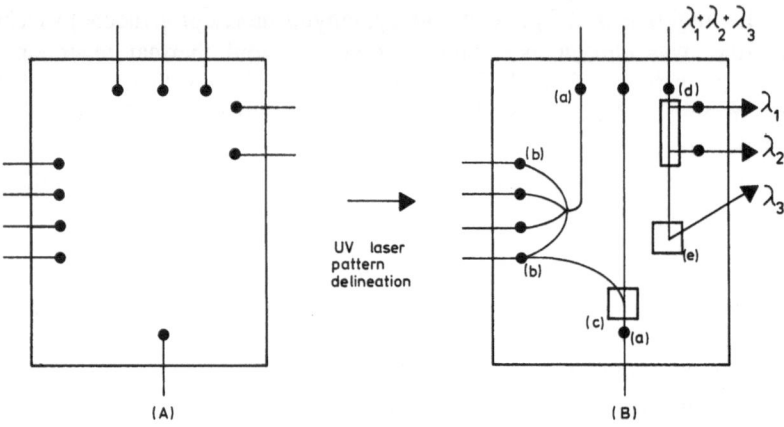

Figure 1.6 Schematic of integrated optical circuits in photochromic polymers. Situation (A) represents a monomode fibre array coupled to a 'latent optical circuit'. Situation (B) represents a UV laser-generated circuit highlighting interconnection to the fibre array and various passive optical components (see text).

sequent optically generated waveguide pattern and highlights a circuit with several features, *viz.* (a) facile wavelength–fibre interconnects, (b) facile generation of complex signal-routing paths, (c) two-dimensional beam splitter (or mirror) for power division and beam coupling, (d) wavelength division demultiplexing (or multiplexing), and (e) out-of-plane decoupling (or coupling). Interferometric laser writing of Bragg diffraction gratings at specific orientations allows refractive index grating generation in planar waveguides. Guided light impinging on such structures under the correct conditions will undergo Bragg diffraction — thus structure (c) in Figure 1.6 may function as a beam splitter or mirror since the level of incident power diffracted is controllable by variation in grating characteristics. The high wavelength selectivity of Bragg diffraction combined with the narrow-wavelength bandwidth (efficient filtering capability) has enabled chirped period gratings (i.e. those with continuously decreasing spacing produced by interfering plane and cylindrical convergent wavefronts) to be used as wavelength division demultiplexers (or multiplexers in reverse) which function as shown in Figure 1.6(d). Gratings with slanted fringes allow structures in planar waveguides capable of coupling light into or out of the guide, as illustrated in Figure 1.6(e).

Many further possibilities for photochromics in integrated optics could exist if active components could be fabricated. In this regard note that certain photochromic spiropyrans, suitably aligned in solution, retain their structure after solvent evaporation, exhibit optical non-linearity and generate the second harmonic (Meredith *et al.*, 1983; Chapter 4, this text). The suitability of such materials for use in active optical devices has not yet been fully described.

Optical bistable effects have been reported in photochromic and related systems. Optical bistability (OB) is the existence of two stable states for one set of optical input conditions. It is a third-order non-linear optical phenomenon, first demonstrated in the mid 1970s (Gibbs *et al.*, 1976, 1980). Two types of OB exist. The first type is the absorptive OB in which a saturable absorber is placed inside and Fabry–Perot etalon configuration tuned off-resonance. The cavity is designed to be an integral number of wavelengths in length with partially reflecting front and rear faces. At low light intensity, destructive interference in the cavity results in low trans-mission through the device. As the input intensity increases, the absorber bleaches and the cavity is tuned to resonance so that a rapid increase in transmitted output can be achieved. Ideally this process is self-sustaining so that one intensity allows a switch-on or up-switching action whilst at a lower intensity the device switches off or down-switches. This characteristic of highly non-linear transmission versus light intensity results in hystersis and bistable behaviour, as shown in Figure 1.7A. The requirements for bleaching in such cases are high in terms of relative absorbtivities between the coloured or less coloured (bleached) forms but can be fulfilled by certain organic dyes and photochromes.

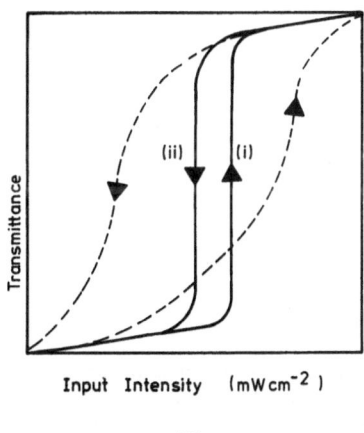

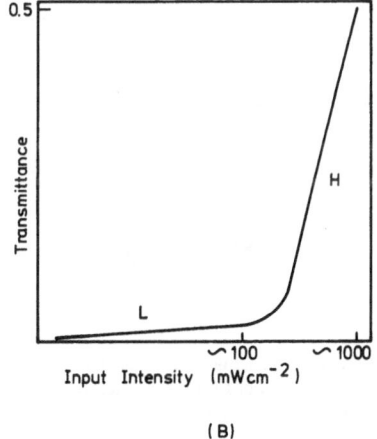

(A)   (B)

Figure 1.7 (a) Schematic representation of idealized hystersis (solid curve) in a Fabry–Perot etalon exhibiting optical bistability; (i) switch-on or up-switching, (ii) switch-off or down-switching — dashed loop is representative of the non-square hystersis exhibited by a photochromic fulgide, PMMA [see Kirkby *et al.* (1985) and text]. (b) Up-switching without bistability in a Fabry–Perot filter employing a liquid solution of photochromic fulgide as saturable absorber: L, low-intensity prethreshold condition; H, high-intensity post-threshold condition (see Mitsuhashi, 1981).

The second type of OB is the dispersive type, which involves tuning the resonant cavity by way of phase shifts in the optically non-linear medium resulting from intensity-dependent refractive index change. This type is more common and is the subject of intense research, particularly in GaAs- and InSb-type materials since the demonstration of true two-beam optical transistor (transphasor) action in which a weak beam can be used to control the characteristics of an intense beam (Gibbs *et al.*, 1980; Miller and Smith 1979).

Because of the refractive index change associated with the absorption change between the coloured and colourless forms of a photochromic species, such materials operate in a complex mode in a Fabry–Perot device configuration, with contributions from both absorptive and dispersive effects influencing cavity performance. This has been elegantly demonstrated by workers at Plessey Research, UK (Kirkby *et al.*, 1985), with solid solutions of E-α-2,5-dimethyl-and 2-methyl, 5-phenyl-3-furyl-ethylidene succinic anhydrides (fulgides) in thin poly(methyl methacrylate) (PMMA) films. In this case equilibrium between photoexcitation and spontaneous decay processes was achieved by simultaneous irradiation of the photochromic film at bleaching and colouring wavelengths, as shown in Figure 1.8.

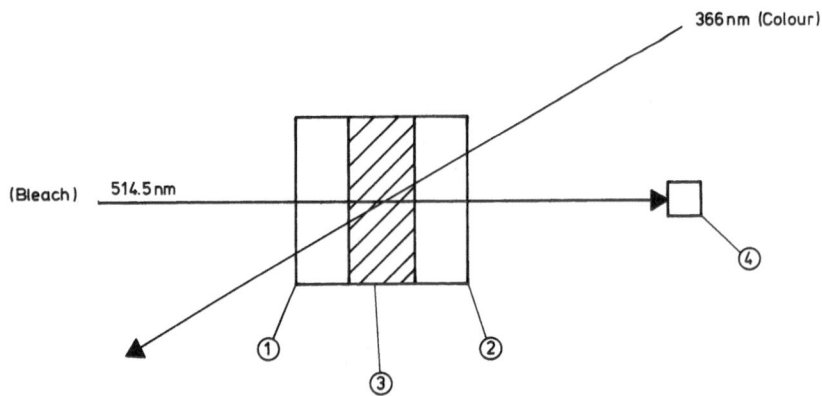

Figure 1.8 Schematic arrangement used by Plessey UK workers to demonstrate optical bistability in photochromic polymer films. 1 and 2, dielectric mirrors of specific reflectivity to bleach and activating wavelengths; 3, 7.5 μm fulgide–PMMA solid solution; 4, detector.

Optical bistability with non-square hystersis (see Figure 1.7A) was observed in plots of 514.5 nm transmission (bleaching wavelength) as a function of the intensity at this wavelength divided by a constant UV irradiance. The effect was seen in the two different fulgide–polymer samples tested, although the up- and down-switching characteristics, maximum transmission and hysteris loop breadths were totally different, highlighting both the need for full materials characterization when using this type of material and the different device operational ranges possible with only subtle variation in photochrome structure. Although the switching transitions were diffuse in these systems, the simplicity of the device and its ease of operation have led to the suggestion that low-cost two-dimensional array fabrication of OB elements for coherent plane-parallel optical processing might be possible (Kirkby *et al.*, 1985).

Up-switching without bistability has also been observed in toluene solutions of the aforementioned (above) fulgide in a Fabry–Perot filter (Mitsuhashi, 1981). Fulgides as a class of photochromes were chosen because of their thermal stability (in coloured forms) and low photochemical fatigue. The photochromic solution was UV pre-exposed then circulated through the cavity.

Mitsuhashi demonstrated image thresholding with this simple saturable Fabry–Perot filter. Thus, when an input transparency with four grey levels was illuminated through the filter at the bleaching wavelength and of low intensity (see region L of Figure 1.7B), the resultant output profile corresponded linearly with four transmission levels. However, when the same transparency was irradiated with intense illumination at the same wavelength in such a way that the power density exceeded the critical power density of the filter (region H, Figure 1.7B) only in the most transparent portion of the input image, the output then became very large relative to the other parts of the input with the result that the four-level input reduced to a single level output, i.e. all subthreshold intensities were filtered out. Similar effects, i.e. up-stitching without bistability, have also been demonstrated with non-photochromic dyes such as specific metalophthalocyanines and cryptocyanines. These materials lack bistability as a result of insufficient bleaching characteristics (Spiller, 1972) but may have applications in optical gating or optical pulse generators with rapid rise times. Saturable absorber dyes (non-photochromic) have also been used in optical signal-processing applications (Lee and Stalker, 1972; Lee, 1974; Stalker and Lee, 1974).

Farhat (1975) has discussed the feasibility of more general non-linear optical data processing with photochromic materials, since these exhibit amplitude transmittance either directly or inversely proportional to the light incident upon them through the use of either initially coloured forms addressed at bleaching wavelengths or initially colourless forms addressed at activating wavelengths respectively.

Various computations (e.g. division) can be realized in parallel by non-linear filtering which may be used to provide new methods for optically encoding information varying in complexity. Subsequent decoding can also be achieved optically through holographic techniques. Image enhancement as a specific example of the types of optical signal processing possible according to Farhat (1975) has been demonstrated in an organic photochromic system by Blažek and Kucharski (1977) with salicylideneaniline, presumably in a solid form, as a material for an adjustable adaptive spatial filter. Such dynamic spatial filters offer advantages over static filters in optical processing in that they allow for greater flexibility and real-time operation. These authors showed that the visibility of low-contrast periodic structures can be substantially improved with the aid of such a filter. Input images consisting of periodic and diffuse non-periodic structures, which may blur the fine details, can be enhanced when the coloured photochromic material is used as a dynamic filter in the Fourier plane of a coherent optical processor and the input data are irradiated at the bleaching wavelength. Grating-like structures formed by sharp peaks with high energy content promote bleaching, whereas the broad spectrum of non-periodic structures does not have sufficient energy to pass through the filter. Such image-enhancement techniques could find outlets in, for example, forensic science (fingerprint examination).

In addition to their ability to function as media capable of preparing complex optical computation on static input data, photochromics can also be used as the active media in optically addressed spatial light modulators (SLMs) which provide re-usable real-time two-dimensional input transducers. Kirkby and Bennion (1986) have described the parameters characterizing photochromic fulgide–polymer SLMs in the input and Fourier planes of the coherent optical processor, indicating that their greatest potential lies in the latter plane by virtue of the high resolution achievable in these systems (≥ 1000 line pair/mm). Their relatively low sensitivity (approx. 1–10 mJ/mm^2) to bleaching imposes constraints in available laser power, frame size and address rate so that no outstanding advantage of the photochromic system as input plane SLMs compared with other types were apparent (Kirkby et al., 1986).

1.5   Conclusions

Optical recording has been used as an exemplary technology to highlight the importance of a strategic approach to photochromic polymer application as well as to stress the important need for an awareness of systems considerations and competitive technologies when seeking to apply these materials. Conceptually, exploitation of the photochromic phenomenon in erasable digital storage applications is perfectly laudable, but from commercial and performance standpoints it is unlikely that photochromic media will ever be used in a practical system design.

Competition in analogue optical recording is fierce, and erasability is not an advantage whereas updatability has niche market potential. Photochromic polymers exploiting fixation techniques may be useful in updatable microform technology.

Polarization holography appears to be a potentially useful technology for optical signal-processing applications. In the organic systems the reversible photochemical transformation of particular photochromes is essential in effecting a high-resolution imaging technique; reversibility is not used in the sense of an erase function, however photochromic anisotropic media may be made erasable. Inorganic photodichroic competitive materials may not equal the resolution capabilities of the organic media, although the former have many attractive features, particularly in processing and encapsulation.

Optical bistable effects take advantage of reversible photochemical phenomena, and applications for photochromics in this area might be feasible in spite of the diffuse optical switching characteristics so far demonstrated in fulgide–polymer systems. The major advances in this area will, however, be made in III–V and II–VI materials.

Regarding integrated optical applications for photochromics it must be said that, whilst the studies to date are elegant and impressive, reversibility in those systems is neither particularly advantageous nor useful so the photochromic phenomenon is again not fully exploited.

Owing to the ease of solid solution preparation (notwithstanding possible difficulties with aggregation), the relatively low 'data rates' for colouration and decolourization, the absolute necessity for reversibility, the exploitation of the high visible absorptivity triggered by ambient UV, the ease of protection of the active matrix by glass and the commercial importance in building energy consumption savings, e.g. in air conditioning, it is easy to understand why glazing-type applications for polymer systems employing modern fatigue-resistant photochromes offering sunlight attenuation control have a more realistic chance of large market exploitation than photonic devices at the present time. The same arguments largely hold true for ophthalmic applications, as will be discussed in Chapter 2.

References

Alexandru, L., Hopper, M., Loutfy, R., Sharp, J. and Vincett, P. (1984). In *Materials for Microlithography*, eds. Thompson, L., Willson, C. and Frechet, J. ASC Symp. Ser. 266, American Chemical Society, Washington DC.

Anderle, K., Birenheide, R., Eich, M. and Wendorff, J.H. (1989) *Makromol. Chem. Rapid Commun.* **10**, 477.

Ando, E., Miyazaki, J., Morimoto, K., Nakahara, H. and Fukuka, K. (1985) *Thin Solid Films* **133**, 21.

Ando, E., Hibino, J., Hashida, T. and Morimoto, K. (1988) *Thin Solid Films* **160**, 279.

Arakawa, S., Kondo, H. and Seto, J. (1984) EPA 0 115 201 to Sony Corp.

Armitage, D. (1981) *J. Appl. Phys.* **52**(7), 4843.

Arnold, G. (1977) *Wiss. Ber. AEG-TELEFUNKEN* **50**(4/5), 128.

Barachevskii, V.A., Lashkov, G.I. and Tsekhomskii, V.A. (1977) *Photochromism and its Applications*. Izd. Khimiya, Moscow (in Russian).

Bennion, I., Hallam, A.G. and Stewart, W.J. (1983) *The Radio and Electronic Engr.* **53**(9), 313.

Bertleson, R.C. (1971) In *Photochromism — Techniques of Chemistry*, Vol. III, ed. Brown, G.H. Wiley Interscience, London Chapter 3.

Blair, H.S. and McArdle, C.B. (1984a) *Polymer* **25**, 999.

Blair, H.S. and McArdle, C.B. (1984b) *Polymer* **25**, 1347.

Blažek, V., Kucharski, M. (1977) *Optical and Quantum Elec.* **9**, 445.

Bordogna, J., Keneman, S.A. and Amodei, J.J. (1972) *RCA Review* **33**, 227.

Borrelli, N.F. and Young, P.L. (1979) *Proc. SPIE* **200**, 51.

Botez, D. (1987) *Laser Focus*, **68**, (March).
Böttcher, H. and Epperlein, J. (1983) *Moderne Photographische Systeme*. VEB Deutscher Verlag fur Grundstaffindustrie, Leipzig.
Bouwhuis, G., Braat, J. Huijser, A., Pasman, J., Van Rosmalen, G., Schouhamer Inmink, K. (1985) *Principles of Optical Disc Systems*. Adam Hilger, Bristol.
Brown, G.H. (ed.) (1971) *Photochromism — Techniques of Chemistry*, Vol. III. Wiley Interscience, London.
Cabrera, I. and Krongauz, V.A. (1987) *Macromolecules* **20**, 2713.
Caimi, F., Casasent, D. and Schneider, I. (1976) *Proc. SPIE* **83**, 25.
Carlson, C.O., Grafton. D.A. and Tauber, A.S. (1962) In *Large Capacity Memory Techniques for Computing Systems*, ed. Yovits, M.C. Macmillan, New York, p. 385.
Chen, M., Rubin, K., Marrello, V., Gerber, U. and Jipson, V. (1985) *Appl. Phys. Letts.* **46(8)**, 734.
Chu, N.Y. (1985) *Proc. SPIE* **562**, 6.
Chu, N.Y. (1986) Eur. Pat. Appl. 195 896 to American Optical Corp.
Chu, N.Y. (1988) *Proc. SPIE* **1016**, 19.
Clarke, W. (1987) *Proc. SPIE* **747**, 128.
Coopers and Lybrand Report (1987) *Information and Image Management: The Industry and the Technologies'* (available from the Association for Information and Image Management, Silver Spring, Maryland, 20910, USA; AIIM Cat. No. D016).
D'Arco, A., Charmet, J.C. and Cloitre, M. (1982) *Revue. Phys. Appl.* **17**, 89.
Delzenne, G.A. (1979) *Adv. Photochem.* **11**, 81.
Dorion, G.H. and Wiebe, A.F. (1970) *Photochromism*. Focal Press, New York.
Dürr, H. (1989) *Angew. Chem. Int. Ed. Engl.* **28**, 413 (Note: *Photochromic Molecules and Systems*, eds Dürr, H. and Bouas-Laurent, H. Elsevier, Amsterdam, in press).
Eckhardt, H., Bose, A. and Krongauz, V.A. (1987) *Polymer* **28**, 1959.
Eich, M. and Wendorff, J.H. (1987) *Makromol. Chem. Rapid Commun.* **8**, 467.
Eich, M. and Wendorff, J.H. (1990) *J. Opt. Soc. Am. B.* **7(8)**, 1428.
Eich, M., Wendorff, J.H., Reck, B. and Ringsdorf, H. (1987) *Makromol. Chem. Rapid Commun.* **8**, 59.
Eisenbach, C.D. (1979) *Photogr. Sci., Eng.* **23**, 183.
Eisenbach, C.D. (1980) *Ber. Bunsenges. Phys. Chem.* **84**, 680.
El'tsov, A.V. (1990) *Organic Photochromes*. Consultants Bureau, New York (English translation of *Organicheskie fotokhromy*, 1987).
Engler, E.M. (1990) *Adv. Materials* **2(4)**, 166.
Farhat, N.H. (1975) *IEEE Trans. Computers* **C-24(4)**, 443.
Fowlis, W.W. (1979) *Opt. Eng.* **18(3)**, 281.
Fukuda, K. and Nakahara, H (1978) *Proceedings of the 7th Internat Congress on Surface Active Substances, Moscow 1976*, **2**, 186.
Furuta, T., Inoue, K. and Mitani, K. (1988) Jap. Pat. 63 33490.
Gibbs, H.M., McCall, S.L. and Venkatesan, T.N.C. (1976) *Phys. Rev. Lett.* **36**, 1135.
Gibbs, H.M., McCall, S.L. and Venkatesan, T.N.C. (1980) *Opt. Eng.* **19**, 463.
Gillespie, R.E. and Lee, S.J. (1979) *Proc SPIE* **169**, 116.
Goodman, H. and Parsons, D.J. (1983) GB 2 142011A, to English Clays Lovering Pochin and Co. Ltd.
Goldberg, M.W. (1979) *Proc. SPIE* **200**, 84.
Goldburt, E., Shvartsman, F. and Krongauz, V.A. (1984) *Macromolecules* **17**, 1876.
Goldburt, E. and Krongauz, V.A. (1986) *Macromolecules* **19**, 246.
Gruler, H., Vilanove, R. and Rondelez, F. (1980) *Phys. Rev. Lett.* **44(a)**, 590.
Gupta, M. and Strome, F. (1985) *Topical Meeting on Optical Data Storage*, WBB1–1.
Hartman, J. and Lind, M. (1987) *Topical Meeting on Optical Data Storage*, Nevada, 155.
Hattori, R., Yagi, T., Yamashita, K. Kagawa, H., Ishii, M., Takamiya, S. and Mitsui, S. (1987) *Proc. SPIE* **740**, 12.
Hauser, C.D. (1983) *J. Appl. Photogr. Engnr.* **9(1)**, 45.
Hawkins, M. and Bowyer, A.G. (1988) EPA 0 279 600 to Courtaulds PLC.
Heller, H.G. (1982) Int. Pat. Appl. WO 83/00568, to Plessey Overseas Ltd.
Hirsberg, Y. (1956) *J. Am. Chem. Soc.* **78**, 2304.
Holden, D.A., Ringsdorf, H., Deblauwe, V. and Smets, G. (1984) *J. Phys. Chem* **88**, 716.
Horrie, K., Tsukamoto, M. and Mita, I. (1985) *Eur. Polym. J.* **21(9)**, 805.

Ichimura, K., Suzuki, Y., Seki, T., Hosoki, A. and Aoki, K. (1988) *Langmuir* **4**, 1214.
Ichimura, K., Suzuki, Y., Seki, T., Kawanishi, Y. and Aoki A. (1989) *Makromol. Chem. Rapid Commun.* **10**, 5.
Lizuka, K. (1971) *J. Appl. Phys.* **42(13)**, 5553.
Lizuka, K. (1972) *Appl. Phys. Lett.* **21(1)**, 33.
Ikeda, T., Horiuchi, S., Karanjit, D.B., Kurihara, S. and Tazuke, S. (1988) *Chem. Lett.* 1679.
Ikeda, T., Horiuchi, S., Karanjit, D.B., Kurihara, S. and Tazuke, S. (1990a) *Macromolecules* **23**, 36.
Ikeda, T., Horiuchi, S., Koranjit, D.B., Kurihara, S. and Tazuke, S. (1990b) *Macromolecules* **23**, 42.
Irie, M. (1987) Jap-Pat. 62/280264.
Irie, M. (1988) *Polymer Preprints* **29(2)**, 215.
Irie, M. and Mohri, M. (1988) *J. Org. Chem.* **53**, 803.
Jacobson, K.I. and Jacobson, R.E. (1976) *Imaging Systems*. Focal Press, London.
Japaridze, K.G. (1979) *Spirochromenes*. Metsniereba, Tbilisi (in Russian).
Jones, P., Darcy, P., Attard, G.S., Jones, W.J. and Williams, G. (1989) *Molec. Phys.*, **67(5)**, 1053.
Kardash, N.S., Krongauz, V.A., Zaitseva, E.L. and Movshovich, A.V. (1974) *Vysokomol. Soed.* **16A**, 390.
Kimura, K., Suzuki, T. and Yokoyama, M. (1989a) *J. Chem. Soc. Chem. Commun.*, 1570.
Kimura, K., Suzuki, T. and Yokoyama, M. (1989b) *Chem. Lett.* 227.
Kirkby, C.J.G., Cush, R. and Bennion, I. (1985) *Opt. Commun.* **56(4)**, 288.
Kirkby, C.J.G. and Bennion, I. (1986) *IEE Proc.* **133**(JI), 98.
Kozak, A. and Williams, G. (1989) *Molec. Phys.* **67(5)**, 1065.
Kryszewski, M. and Nadolski, B. (1977) *Pure Appl. Chem.* **49**, 511.
Kurihara, S., Ikeda, T. and Tazuke, S. (1988) *Jpn. J. Appl. Phys.* **27(10)**, L 1791.
Kurtz, R.L. and Owen, R.B. (1975) *Opt. Eng.* **14**, 393.
Kvasnikov, E.D., Shatun, V.V. and Barachevskii, V.A. (1973) *Sov. J. Quantum Electronics* **2(4)**, 356.
Lamarre, L. and Sung, C.S.P. (1983) *Macromolecules* **16**, 1729.
Lampert. C.M. and Granqvist, C.G. (eds.) (1990) *Large Area Chromogenics: Materials and Devices for Transmittance Control*, SPIE Vol. IS4, SPIE — International Society for Optical Engineering, Washington DC, p. 2.
Lee, S.H. and Stalker, K.T. (1972) *J. Opt. Soc. Am.* **62**, 1366.
Lee, S.H. (1974) *Opt. Eng.* **13**, 196.
Lo, D.S. and Honebrink, R.W. (1972) *Proc SPIE* **4**, 239.
McArdle, C.B., Blair, H.S., Barraud, A. and Ruaudel-Teixier, A. (1983) *Thin Solid Films* **99**, 181.
McArdle, C.B. (1989) In *Side Chain Liquid Crystal Polymers*, ed. McArdle, C.B. Blackie and Son, Glasgow, p. 357.
McArdle, C.B. and Blair, H.S. (1984) *Colloid Polym. Sci.* **262**, 481.
Meredith, G.R., Williams, D.J., Fishman, S.N., Goldburt, E.S. and Krongauz, V.A. (1983) In *Nonlinear Optical Properties of Organic and Polymeric Materials*, ACS Symposium Series 233. American Chemical Society, Washington DC, Chapter 6, p. 135.
Miller, D.A. and Smith, S.D. (1979) *Opt. Commun.* **31**, 101.
Mitsuhashi, Y. (1981) *Opt. Lett.* **6(3)**, 111.
Miyauchi, A., Ohnishi, M., Noguchi, M. and Washizawa, Y. (1983) *J. Appl. Photogr. Engnr.* **9(1)**, 7.
Miyazaki, J. and Ando, E. (1986) EPA 0 230 024 to Matsushita Electric Industrial Co. Ltd.
Mlles, D.G., Lamb, P.D., Rhee, K.W. and Johnson, C.S. (1983) *J. Phys. Chem.* **87**, 4815.
Möbius, D., Bücher, H., Khun, H. and Sondermann, J. (1969) *Ber. Busenges. Phys. Chem.* **73**, 845.
Morgan, D.A. (1979) *Proc. SPIE* **169**, 105.
Morin, M., LeBlanc, R.M. and Gruda, I. (1980) *Can. J. Chem.* **58**, 2038.
Nagamura, T., Isoda, Y., Sakai, K. and Ogawa, T. (1990) *J. Chem. Soc. Chem. Commun.*, 703.
Nakagiri, T., Nishimura, Y., Sakai, K., Tomita, Y., Eguchi, T. and Saito, K. (1986a) Jap. Pat. 61/175084 A2 to Canon KK.

Nakagiri, T., Nishimura, Y., Sakai, K., Tomita, Y., Eguchi, T. and Saito, K. (1986b) Jap. Pat. 61/175087 A2 to Canon KK.
Nakagiri, T., Nishimura, Y., Sakai, K., Tomita, Y., Eguchi, T. and Saito, K. (1986c) Jap. Pat. 61/176923 A2 to Canon KK.
Nakagiri, T., Nishimura, Y., Sakai, K., Tomita, Y., Eguchi, T. and Saito, K. (1986d) Jap. Pat. 61/175082 A2.
Ogura, K., Hirabayashi, A., Ueijima, A. and Nakamura, K. (1982) *Jpn. J. Appl. Phys.* **21**, 969.
Ohno, S. (1975) *J. Photogr. Sci. Engr* **19(5)**, 287.
Ortler, R., Brauchle, C., Miller, A. and Riepl, G. (1989) *Makromol. Chem. Rapid Commun.* **10**, 189.
Palm, C.S., Weaver, S.E., Binns, B.W. and Mercer, C.C. (1979) *Proc. SPIE* **200**, 27.
Polymeropoulous, E. and Mobius, D. (1979) *Ber. Busenges. Phys. Chem.* **12**, 1215.
Plumadore, J.D. and Spiteri, C.B. (1979) *Proc SPIE* **169**, 112.
Reichmanis, E. and O'Donnell, J.H. (eds.) (1989) *The Effects of Radiation on High Technology Polymers*, ACS Symposium Series 381. American Chemical Society, Washington DC.
Reid, P.I. and Waters, B.R. (1983) Evr. Pat. Appl. 0 070 631, to English Clays Lovering Pochin and Co. Ltd.
Richert, R. (1988) *Macromolecules* **21**, 923.
Richert, R. and Baessler, H. (1985) *Chem. Phys. Lett.* **116**, 302.
Ritzau, T. (1987) *Proc SPIE* **747**, 119.
Shaffert, R.M. (1975) *Electrophotography.* Wiley and Sons, New York.
Shankoff, T.A. (1969) *Appl. Opt.* **8**, 2282.
Shen, S. Torkelson, J.M. Elbert, J.E. and Lewis, F.D. (1990) *Makromol. Chemie* **191(10)**, 2367.
Shieh, H. and Kryder, M. (1987) *J. Appl. Phys.* **61(3)**, 1108.
Smets, G. (1975) *J. Polym. Sci., Polym. Chem. Ed.* **13**, 2223.
Smets, G. (1983) *Adv. Polym. Sci* **50**, 17.
Spiller, E. (1972) *J. Appl. Phys.* **43(4)**, 1673.
Stalker, K.T. and Lee, S.H. (1974) *J. Opt. Soc. Am.* **64**, 545.
Sugiyama, Y. and Sakai, W. (1987) Jap. Pat. 62 252 496.
Sung, C.S.P., Gould, I.R. and Turro, N.J. (1984) *Macromolecules* **17**, 1447.
Tachibana, H., Nakamura, T., Matsumoto, M., Komizu, H., Manda, E., Niino, H., Yube, A. and Kawabata, Y. (1989) *J. Am. Chem. Soc.* **111**, 3080.
Tamura, S. and Seto, N. (1986) Jap. Pat. 61/28939 A2 to Sony Corp.
Tazuke, S., Kurihara, S., Yamaguchi, H. and Ikeda, T. (1987a) *J. Phys. Chem.* **91**, 249.
Tazuke, S., Kurihara, S. and Ikeda, T. (1987b) *Chem. Letts.* 911.
Todorov, T., Nikolova, L. and Tomova, N. (1984) *Appl. Opt.* **23(23)**, 4309.
Todorov, T., Nikolova, L., Stoyanova, K. and Tomova, N. (1985) *Appl. Opt.* **24(6)**, 785.
Todorov, T., Nikolova, L., Tomova, N. and Dragostinova, V. (1986) *IEEE J. Quant. Elec.* **QE-22(8)**, 1262.
Tomlinson, W.J. (1984) *Appl. Opt.* **23(24)**, 4609.
Tomlinson, W.J. and Chandross, E.A. (1979) *Adv. Photochem.* **12**, 201.
Tsai, F.-J., Torkelson, J.M., Lewis, F.D. and Holman, B. (1990) *Macromolecules* **23**, 1487.
Ueda, Y. (1987) *Jpn. J. Electr. Eng.*, 51.
Urita, K. (1987a) *Jpn. Semicond. Tech. Repts.* **2(4)**, 36.
Urita, K. (1987b) *Jpn. J. Electr. Eng.*, 31.
Victor, J.G. and Torkelson, J.M. (1987a) *Macromolecules* **20**, 2241.
Victor, J.G. and Torkelson, J.M. (1987b) *Macromolecules* **20**, 2951.
Victor, J.G. and Torkelson, J.M. (1988) *Macromolecules* **21**, 3490.
Vilanove, R., Hervert, H., Gruler, H. and Rondelez, F. (1983) *Macromolecules* **16**, 825.
Weigert, F. (1919) *Verh. Dtsch. physik. Ges.* **21**, 479.
Whitten, D.G. (1979) *Angew. Chem. Int. Ed. Engl.* **18**, 440.
Wilson, A.E.J. (1984) *Phys. Technol.* **15**, 232.
Wu, C.-K. (1979) *Proc SPIE* **200**, 39.
Wu, G. and Yunfei, M. (1988) *Kexue, Tongbao* **33(9)**, 742 (in English).
Yamaguchi, T. (1987) *Jpn. J. Electr. Eng.*, 54.
Yitzchaik, S., Cabreca, I. and Buchholtz, F. (1990) *Macromolecules* **23**, 707.
Yoshida, T., Morinaka, A. and Funakoshi, N. (1987) JP 62/254144 AZ to NTT Co.
Yu, W.-C. and Sung, C.S.P. (1988) *Macromolecules* **21**, 365.

2 Spiroxazines and their use in photochromic lenses

J.C. CRANO, W.S. KWAK and C.N. WELCH

2.1 Introduction

The discovery of a system based on silver halide crystallites suspended in an inorganic glass matrix resulted in the development of the first commercially acceptable photochromic ophthalmic lenses (Armistead and Stookey, 1964, 1965). The convenience of having lenses that darken automatically upon exposure to sunlight has proven to be appealing to spectacle-wearers. An estimated 10 million pairs of silver-based photochromic prescription lenses were dispensed in the United States in 1989 (Optical Manufacturers Association, 1990).

Another trend within the ophthalmic industry over the last several years has been the growth of plastic lenses. Again taking the United States market for illustration, the percentage of prescription eyewear made of plastic has reached 70% in 1989 (Optical Manufacturers Association, 1990). With this increasing market penetration of plastic lenses, the desire for plastic photochromic ophthalmic lenses has also increased.

The discovery of indolinospironaphthoxazines as a family of photochromic compounds with inherent fatigue resistance led to early attempts at commercialization of organic systems for the prescription lens market. The first references to photochromic spiroxazines appeared in 1970. Both Ono and Osada (1970) and Arnold and Vollmer (1970) disclosed indolinospironaphthoxazines derived from 1-nitroso-2-naphthol. Subsequently, Hovey *et al.* (1980) disclosed spironaphthoxazines having selected substitution on the naphthoxazine ring system. More recently, Kwak and Hurditch (1987) described replacing the spironaphthoxazine ring system with spiropyridobenzoxazine.

2.2 Chemistry of spiroxazines, spiropyrans and related organic photochromic compounds

2.2.1 Nomenclature

Four representative spiro-organic photochromic compounds are shown in

Figure 2.1 with their names and numbering system. For each compound two names are given. The preferred names are those according to the IUPAC rules of organic nomenclature which are used in the Chemical Abstract index for the period 1967–1971. Current Chemical Abstract index names for these compounds are also given. However, it is often more convenient to use abbreviations rather than full names. The commonly accepted abbreviations SP (spiropyran), SO (spiroxazines) and MC (merocyanines) have been used throughout this discussion. Furthermore, SP and SO designations are used for the colorless spiropyrans and spiroxazines respectively. On the other hand, open-colored species of the photochromics have merocyanine structures and, therefore, such structures are represented by MC for all classes of the spiro-organic photochromics. For example, the abbreviation of the compound shown in Figure 2.1(a) is 6-nitro-BIPS because it is a derivative of the simplest member of the class BIPS(1′,3′,3′-trimethylspiro-[2H-1-benzopyran-2,2′-indoline]) with a nitro group at the 6 position. Likewise, a compound for which the abbreviation is 3-ethyl-9′-methoxy-NISO is depicted in Figure 2.1(c). Notice that the indoline part (or left-hand side) of

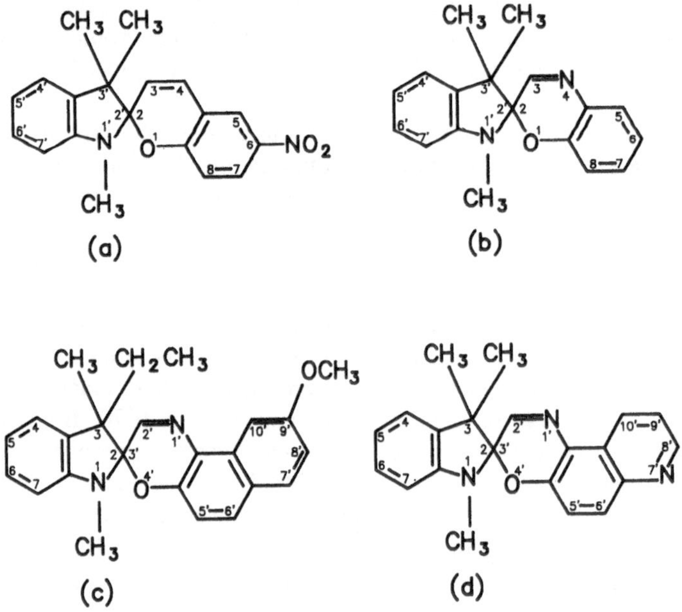

Figure 2.1 Structure and nomenclature of four spirophotochromics. (a) IUPAC, 6-nitro-1′,3′,3′-trimethylspiro[2H-1-benzopyran-2,2′-indoline]; CA index, 1′,3′-dihydro-6-nitro-1′,3′,3′-trimethylspiro[2H-1-benzopyran-2,2′-[2H]indole]; abbreviation, 6-nitro-BIPS. (b) IUPAC, 1′,3′,3′-trimethylspiro[2H-1,4-benzoxazine-2,2′-indoline]; CA index, 1′,3′-dihydro-1′,3′,3′-trimethylspiro[2H-1,4-benzoxazine-2,2′[2H] indole]; abbreviation BISO. (c) IUPAC, 3-ethyl-9′-methoxy-1,3-dimethylspiro[indoline-2,3′-[3H]naphth-[2,1-b][1,4]oxazine]; CA index, 1,3-dihydro-3-ethyl-9′-methoxy-1,3-dimethylspiro[2H-indole-2,3′-[3H]naphth[2,1-b][1,4]oxazine]; abbreviation 3-ethyl-9′-methoxy NISO. (d) IUPAC, 1,3,3-trimethylspiro[indoline-2,3′-[3H]pyrido[3,2-f]-[1,4]benzoxazine]; CA index, 1,3-dihydro-1,3,3-trimethylspiro[2H-indole-2,3′-[3H]pyrido-[3,2-f][1,4]benzoxazine]; abbreviation QISO.

the BIPS and BISO molecules are numbered with primes, while NISO and QISO classes are designated with primes on the oxazine (or right-hand) part of the molecules. This difference is a result of the naming of the fundamental ring systems according to the IUPAC rules of organic nomenclature.

2.2.2 Synthesis

An extensive number of organic photochromic compounds containing a spiro carbon has been well documented (Bertelson, 1971). In general, spiropyrans, spiroxazines and structurally related organic photochromics are synthesized by a thermal condensation reaction of the corresponding alkylidene heterocycle or its conjugate acid with *ortho*-hydroxyformyl or *ortho*-hydroxynitroso aromatic derivatives. They form readily in most polar organic solvents at or near the reflux temperatures and can be purified by recrystallization or column chromatography. Reaction schemes for SOs and SPs are shown in eqns (1) and (2) respectively. Bertelson has also elucidated other synthetic approaches for different types of spiro carbon-containing photochromics. Those methods were used for compounds with either more than one spiro carbon atom or for which intermediates were not readily available.

Of all spiro-organic photochromics, probably 6-nitro-1′,3′,3′-trimethylspiro [2H-1-benzopyran-2,2′-indoline](6-nitro-BIPS) has been most extensively studied. Therefore, synthesis of 6-nitro-BIPS is described as an example. It is prepared by a slightly modified literature method (Inoue et al., 1968; Sivadjian 1968). A Fischer base, 1,3,3-trimethyl-2-methyleneindoline (3.5 g; 0.02 mol), was dissolved in 40 ml of hot absolute ethanol. To this was added 3.35 g of 2-hydroxy-4-nitrobenzaldehyde (0.02 mol) in small portions in 10–15 min. After the dark pink–brown solution had been refluxed for 2 h, the reaction mixture was allowed to evaporate to a small volume in an air draft. The fine powder was collected by filtration, washed with absolute ethanol and then air-dried. This was recrystallized from boiling n-hexane with a small amount of activated charcoal. A pale yellow microcrystalline powder was obtained.

In the last several years a relatively large number of spiroxazines (NISO) and spiropyridoxazines (QISO) have been reported. They can also be prepared in a similar way. For example, 1,3,3-trimethyl-spiro[indoline-2,3′-[3H]naphth[2,1-b][1,4]oxazine](NISO) was formed when stoichiometric amounts of the Fischer base and 1-nitroso-2-naphthol were refluxed in alcohol or toulene for 2–4 h (Ono and Osada, 1970). The product was collected by filtration, washed with alcohol and air-dried. It was recrystallized from boiling n-hexane with a small amount of activated charcoal. A pale-greenish powder was obtained. QISO and BISO derivatives can be prepared by similar methods (Kwak and Hurditch, 1987; Kwak 1989; Kwak and Chen, 1989a).

2.2.3 Photochemistry of the spirophotochromics

2.2.3.1 Photocoloration and thermal and photobleaching. For historical reasons alone, it is appropriate to discuss the photochemistry of photochromic spiropyrans first, especially 6-nitro-BIPS. Pioneering work by the Weizmann Institute group (Hirshberg, 1950; Hirshberg and Fisher, 1953; Heiligman-Rim *et al.*, 1962; Bercovici, 1969) led to the suggestion that spiropyrans are colored because of the formation of merocyanine-like structures caused by UV light absorption. Later, Becker and his group demonstrated, based on spectral data and independent syntheses, that the open-colored form of spiropyrans indeed possesses the merocyanine-like structure (Kolc and Becker, 1967). In the absence of an activation source, the merocyanines bleach back to the colorless closed spiro structures thermally and or photochemically.

$$\text{SP or SO} \xleftarrow[\quad h\nu_2 \text{ and/or } kT \quad]{\quad h\nu_1 \quad} \text{Mc}$$

Colorless   Colored

The most interesting and fundamental question of photochromism is how the colorless SP and SO molecules become colored when they are irradiated with UV light. The next question is how they bleach back to the colorless state by heat and/or light. Also included in this quest is the important observation that the same molecule behaves differently in different environments. These aspects are rather well understood for many SP and SO compounds through the results of laser flash photolysis experiments and theoretical calculations. Because of the nature of this chapter, only the very basic and elementary mechanisms are briefly discussed.

The SP and SO molecules consist of two halves of heterocyclic fused rings connected at the spiro carbon. However, these two near-planar aromatic parts are electronically isolated from each other because of the orthogonality of both the topological system and the pi-electron wave functions (Tyer and

Becker, 1970). The electronic transitions of such a spiro compound are essentially localized on each half of the molecule. The lowest electronic transition of most of the SP and SO molecules occurs in the (near) UV region of the electromagnetic spectrum and, therefore, the compounds are (almost) colorless or pale yellow. When an SP or SO compound absorbs UV light, the relatively weak spiro carbon–oxygen bond undergoes a heterolytic cleavage to form a merocyanine-like structure (Heiligman-Rim, 1961). Unlike the closed form of the molecules, in the MC structure the two parts of the fused ring systems are (nearly) coplanar. The pi-electron cloud in the 'open' form can extend from the indoline part through the methine (or azamethine) bridge to the pyran (or oxazine) part of the molecule. The first electronic transition of such a photomerocyanine molecule occurs in the visible region, thus appearing colored (Figure 2.2).

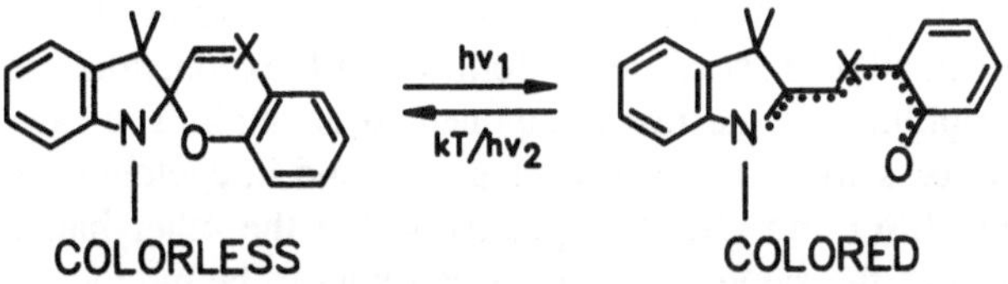

Figure 2.2 Structures of colorless SP or SO and colored MC (Heiligman–Rim, 1961).

A recent flash photolysis study of NISO and QISO has shown that the spiro carbon to oxygen bond rupture takes place during the laser pulse (pulse width at half-width: 2 ns) (Kellmann et al., 1989). It was also found that two or more of the stable geometric isomers reach an equilibrium state very quickly in both toluene and methanol solutions at 298 K (see below). These results are in agreement with the picosecond laser studies (Schneider 1987a, b) and a Raman study (Kluter et al., 1985). Another observation that has been made is that formation of the MC is independent of oxygen. This indicates that the spiro carbon to oxygen bond breaking occurs only in the excited singlet state of the closed form of NISO and QISO. There is no evidence indicative of the participation of a triplet state in the coloration process of NISO or QISO. Similar conclusions have been reported recently for the parent NISO (Bohne et al., 1990). (The 'A' represents the colorless SP or SO and the 'Bs' are the stereoisomers of the thermally stable colored transoids.)

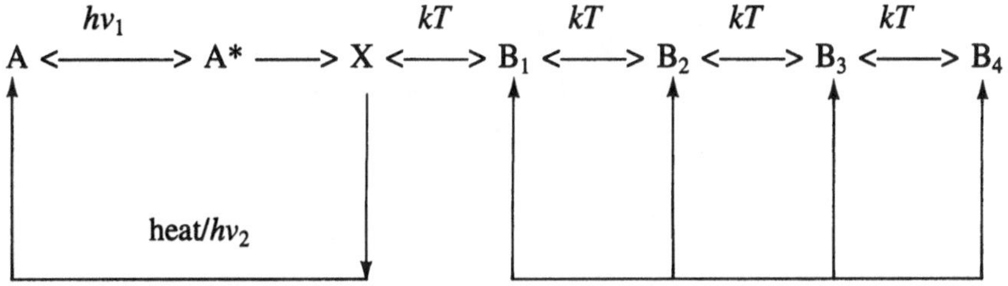

The photomerocyanines bleach back to the colorless closed form by thermal and/or photolytic processes. Chu (1985a) has reported that in ethanol the thermal bleaching rate of the MC of NISO increases about three times for every 10°C temperature increase in the 0–30°C range. The thermal bleaching rate constants at 10, 20 and 30°C were 0.0798, 0.212 and 0.61/s respectively. The activation energy for MC of the parent NISO for bleaching was 19.7 kcal/mol. It is interesting to note that the energy difference between the ground states of the colorless and colored forms of the parent NISO was only 4.2 kcal/mol. Because of the fast decay of MC at higher temperatures, saturation absorbance of an MC at the steady state decreases quite dramatically as temperature increases. The relationships between temperature, thermal bleach rate and steady-state coloration are shown in Figures 2.3 and 2.4 respectively (Chu, 1983a).

Thermal ring closure mechanisms of MC of NISO and QISO were found to be different in polar and non-polar solvents (Kellmann et al., 1989). In non-polar solvents the thermal decay followed a biexponential pathway, although the amplitude of the fast component was small. The rate constants of the slow process were 0.75 and 0.71 per second in cyclohexane at 298 K for the parent NISO and QISO respectively. On the other hand, the ring-closure kinetics of the same molecules was found to be monoexponential in

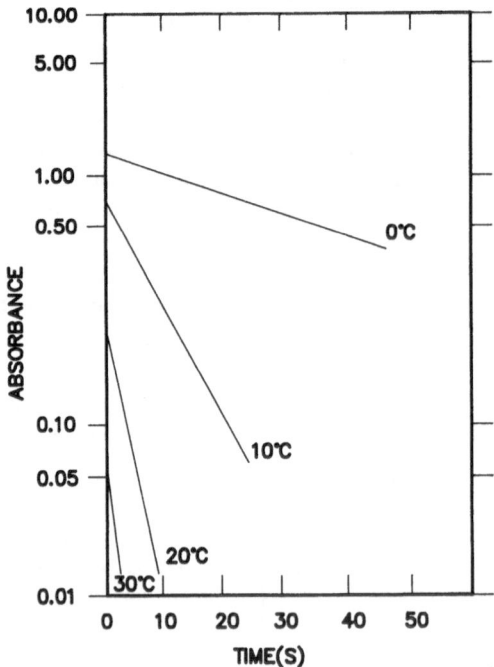

Figure 2.3 Thermal decay of the colored form of NISO. (After Chu, 1983a.)

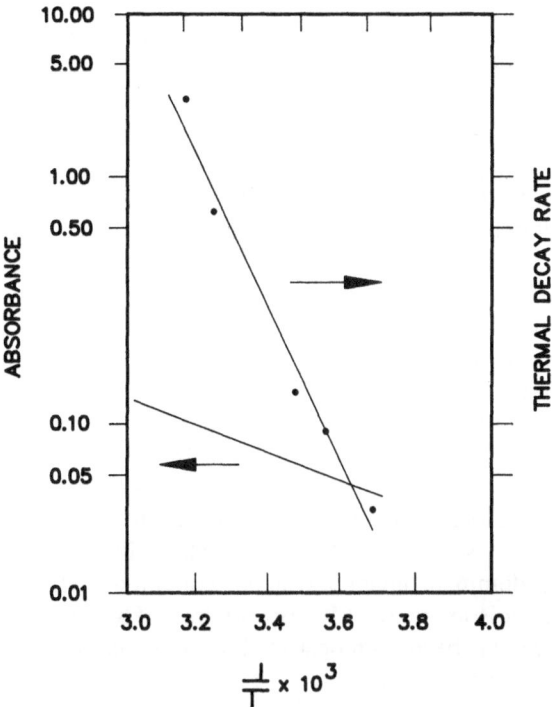

Figure 2.4 Thermal decay rate and equilibrium absorbance versus $1/T$. (After Chu, 1983a.)

polar solvents. The rate constants were 1.85 and 0.82 per second for NISO and QISO respectively in acetonitrile at 298 K. A recent study using a two-laser flash technique has shown that the photolytic decay of MC of the parent NISO follows different pathways in polar and non-polar solvents. The involvement of an unspecified species 'X' was suggested during the photobleaching in polar solvents (Bohne *et al.*, 1990).

2.2.3.2 Nature of the colored and colorless forms. Direct evidence for structures of both the closed and open forms has been obtained by X-ray crystallographic structure determination of several spiroxazines and spiropyrans. The results have shown that the spiro carbon to oxygen bonds of the colorless SP and SO molecules are 0.01–0.05 Å longer than those in a number of oxazines and pyrans (Clegg *et al.*, 1987, 1990; Aoto *et al.*, 1989). The longer spiro carbon to pyran (or oxazine) oxygen is a result of the strained five-membered ring that includes the spiro carbon. This is in agreement with the rationale for the rupture of spiro carbon to oxygen bond by UV light absorption by the photochromic molecules. Direct crystallographic evidence for the merocyanine structure of the colored species was reported for a number of 6-nitro-BIPS derivatives by Aldoshin and Atov-myan (1987). For example, the two fused rings of both ends of the open form of 8-bromo-6-nitro-BIPS were found to deviate slightly from coplanar as a result of rotation about two of the methine bonds by 11.7 and 4.6° respectively in the dark-purple crystals. Such a deviation from coplanar configuration might be required by the intramolecular steric constraints and/or the packing mode in the crystals. The similarity of the colors of the open forms in both solution and crystal suggests that both of them should have similar structures. None of the merocyanine forms of SO compounds has been examined by X-ray diffraction.

The open colored MC can exist in eight stereoisomeric forms. Among them, the four transoids are energetically more favorable because of the lack of steric hindrance compared with the cisoids (see Figure 2.5). At the same time, they have a linear structure for more effective pi-electron conjugation than the angled cisoids. All of the isomers are thermally interconvertible from one another by no more than several kcal/mol (Aoto et al., 1989). Nonetheless, there are significant differences in thermal stability of the *trans* isomers (a)–(d), and such a difference makes one or two species predominant under the given conditions. The interconvertibility of the stereoisomers of MC of 6-nitro-BIPS by small thermal energies has elegantly been demonstrated at low temperatures by the Weizmann Institute chemists (Bercovici, 1969). Many MC spectra of SP and SO derivatives indicate that more than one isomer exists in equilibrium at room temperature in solutions (Heiligman-Rim et al., 1962; Kellmann et al. 1989).

The open MC molecules have exhibited several characteristics that can be explained by three mesomeric forms as shown in Figure 2.6. They are

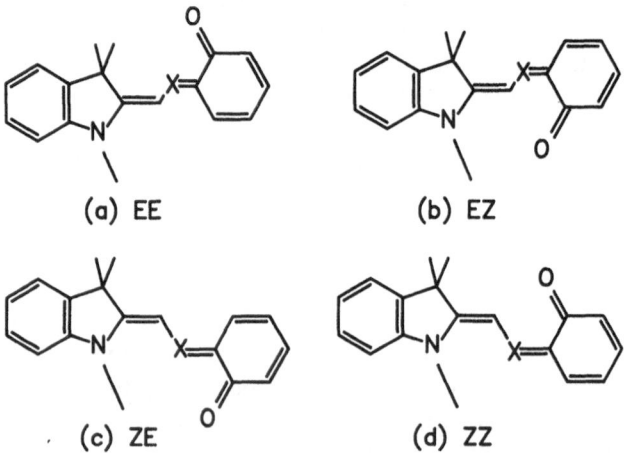

(a) EE   (b) EZ

(c) ZE   (d) ZZ

Figure 2.5 Four geometrical isomers of the transoid MC.

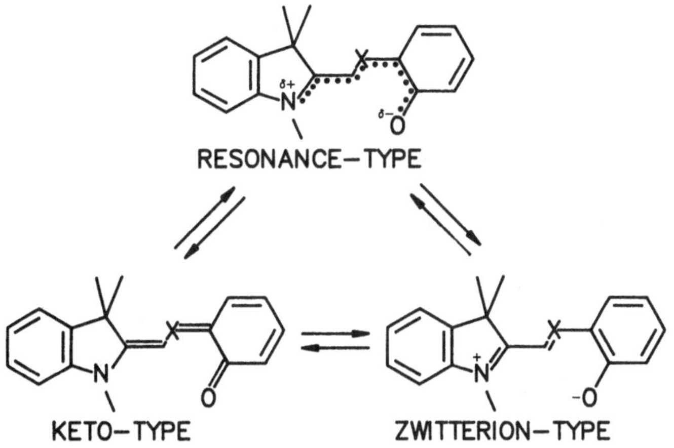

RESONANCE−TYPE

KETO−TYPE   ZWITTERION−TYPE

Figure 2.6   Three possible electronic structures of the colored MC.

the quinoidal, zwitterionic and the resonance hybrid structures (Bertelson, 1971). All experimental data and molecular orbital calculations suggest that the resonance hybrid is the most accurate representation of the colored molecules. Nonetheless, the degree of electron delocalization along the methine (SP) or azamethine (for SO) chain is quite different depending upon the individual molecule. Therefore, some MC species behave as if they possess a quinoidal structure and others show more zwitterionic characteristics. More specifically, thermodynamic and kinetic behaviors of most of the photocolored MC forms of NISO and QISO have been found to be more consistent with the quinoidal than the zwitterionic structure (Kellmann *et al.*, 1989). Geometry optimization of the open form of NISO by MNDO calculation also suggests that the open form of the parent NISO has a nonplanar quinoid structure (Aoto *et al.*, 1989). On the other hand, the MC form of 6-nitro-BIPS exhibits spectroscopic characteristics that are more consistent with the zwitterionic structure (Bertelson, 1971).

Another interesting property of the MC species is solvatochromism. The photoactivated spectrum of the parent NISO has shown an absorption maximum at 590 nm with a shoulder near 570 nm in toluene. They are shifted to 610 and 580 nm respectively in methanol (Kellmann *et al.*, 1989). The observed positive solvatochromism is characteristic of a compound having a weakly polar ground state. MNDO calculations have shown that the ground state and the first singlet excited state of the parent NISO have dipole moments of 3.8 and 5.4 debyes respectively (Aoto *et al.*, 1989). Both the spectroscopic data and semiempirical calculation are more consistent with a quinoidal structure for the MC of this photochromic compound. The parent QISO was also reported to show a similar positive solvatochromism

(Kellmann *et al.*, 1989). Other spectroscopic properties of the photoactivated species of SP compounds have been studied in relation to the solvent properties (Flannery, 1968).

2.2:3.3 Quantum yield and photo- and thermal degradation. Quantum yields for the forward coloration reaction of NISO and QISO have been measured to be relatively small (Kellmann *et al.*, 1989). This can be explained by the observations that the MCs of NISO and QISO are formed only directly from the first excited singlet state of the closed form. This process should compete with the internal conversion of the first singlet excited state to the ground state and/or from the cisoid of the open form to the ground state. This is consistent with the result that no luminescence was observed for NISO in ethanol (Chu, 1983a). A recent study by two-laser flash photolysis also reported that a triplet state is not involved in the coloration mechanism of NISO in solutions (Bohne *et al.*, 1990).

One of the major deficient properties of spiropyrans (SP) for many potential applications is their short life-cycle because of the extensive thermal and photochemical decomposition (Bertelson, 1971). On the contrary, NISO and QISO derivatives have been reported to possess good fatigue resistance (Chu, 1985a; Kellmann *et al.*, 1989, Bohne *et al.*, 1990). For example, Chu (1985a) has reported that the quantum yield for the photodecomposition of the parent NISO was 0.0002 in ethanol to 366 nm light of a mercury lamp. This is about 1000 times smaller than those reported for most SPs. A similar study by Bohne *et al.* (1990) using 308-nm laser irradiation showed a photodegradation quantum yield of NISO to be less than 0.05. The parent NISO deteriorated more extensively in non-polar than polar solvents. No photodegradation mechanism or fatigue product analysis has been reported for NISO or QISO.

2.3 Required properties of photochromic systems for eyewear

It has been suggested that a commercial prescription ophthalmic photochromic lens should meet several minimal requirements to be accepted by the consumer. The bleached hue should be near-colorless for a non-sunlens application and a pleasing brownish or grayish colour for the sunlens application. The activated hue should have wide consumer acceptance. The rates of both UV activation and thermal bleaching should be sufficiently fast that density changes are readily perceptible to the wearer. The extent of the fully developed density, or response, must offer real glare protection for the wearer and be functional over a broad range of exposure conditions. Lastly, the photochromic effect must not noticeably change over the normal eyewear lifetime.

2.3.1 General

The human eye exists in a complex variable world and is required to accommodate a wide range of lighting conditions. The amount of light that enters the eye is controlled by the iris whose muscles modulate the opening called the pupil, through which the light must pass. The size of the pupil is automatically adjusted to varying degrees of light intensity, being a very small diameter with high light intensities and progressively larger with lower intensities. The cornea and the lens then work in unison to focus the light ray image upon the retina. In this respect, the eye is very similar to a camera in that the amount of light entering the camera and the focusing of the light rays onto the film are controlled. However, a normal eye can focus both near and far-distant objects upon the retina without changing the lens to retina distance, while a camera must adjust the lens to film distance. This phenomenon of the eye's lens is called 'accommodation'. The accommodation factor will be discussed further in section 2.6.3.1.

The eye is very sensitive to the intensity of light and has two levels of light detection. The light sensors responsible are called 'rods' and 'cones'. The 'rods' are sensitive only to low intensities of light and become inactivated at higher intensities. They are basically used for night vision and require long periods of time (20–60 min) to become fully developed for maximum utilization. On the other hand, the 'cones' respond to changes in light intensities much more quickly and become fully developed for maximum utilization within 5–10 min. It is also a fact that these 'rods' and 'cones' contain visual pigments and are more sensitive to specific wavelengths of light; in other words, they exhibit spectral selectivity (color vision). This sensitivity of 'rods' and 'cones' as a function of wavelength is best illustrated in Figure 2.7. It can be concluded as expected that the 'rods' are much more sensitive than the 'cones' with a maximum sensitivity in the blue–green region (scotopic, with the wavelength of maximum absorption at approximately 510 nm), while the 'cones' have a maximum sensitivity in the yellow–green region (photopic, with the wavelength of maximum absorption at approximately 555 nm). The 'rods' yield only achromatic or neutral color perceptions — white, gray and black — while the 'cones' give perception of chromatic colors. In the red region, they both respond approximately equally, and with significant attenuation.

There are three types of 'cones' which contain different pigments. Each pigment absorbs light over the major part of the visible spectrum; however, the spectral selectivity of each specific pigment enables the eye to exhibit color vision (Rossotti, 1983). All three pigments contribute to color vision more or less equally at the shorter wavelengths, while only one of the pigments basically contributes at the longer wavelengths.

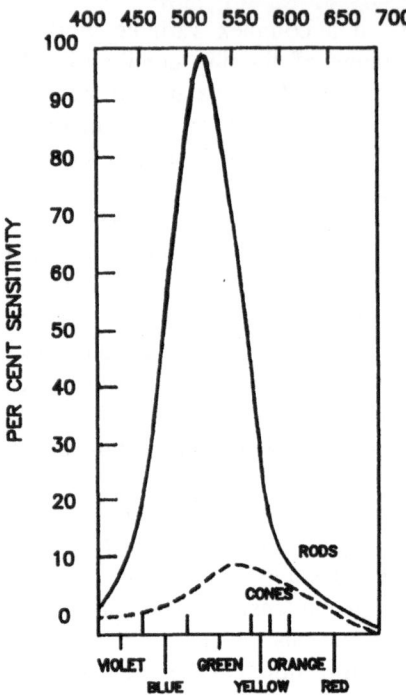

Figure 2.7 Comparison of the relative sensitivity of the rods and cones in the eye to visible light (adapted and redrawn from Sears, 1949.)

2.3.2 Absorption spectra of unactivated and activated states

A photochromic lens would operate in the mode of functioning as a variable filter for the eye. Basically, under low light intensity conditions, such as indoors, the lens would exhibit a high transmission, the preferred range being 80–92%. When solar-exposed, the photochromic would be activated with a subsequent reduction in transmission, and thus the eye would receive a significant reduction in light intensity.

The basic function of a photochromic lens is to protect the eye from unwanted light or 'glare'. Glare can be classified into three catagories: scenic, scattered and reflected. Scenic glare is due to excessively high intensities of light, such as that which is experienced on a typical summer day or at the beach. Scattered glare results when light is reflected and refracted as a result of the presence of dust particles, gases and water vapor. On a clear day, the short wavelengths (blues) are preferentially scattered, which results in a sky with a blue hue. On dustier and more humid days, the sky actually becomes whiter due to the scattering of the longer (red) wavelengths of light. Reflected or spectral glare is an enhanced brightness which results when sunlight strikes a horizontal shiny surface at the proper angle. A photochromic lens can reduce scenic and scattered glare, but only a special polarized lens can reduce or alter reflected glare.

The reduction of scenic glare can best be achieved with lenses that are broad-band spectral filters, such as those with gray and brown hues. The spectra of such broad-band filters are shown in Figure 2.8, simulated by tinted gray and brown lenses cast from CR-39® monomer. Such broad-band lenses attenuate all wavelengths of the incoming light rays, with the gray attenuating more uniformly. As a result, the color vision of the eye is not notably distorted and *full color* recognition is enjoyed by the lens-wearer. These types of lenses are most frequently incorporated into sunglasses. The brown and gray tints, in addition to being visually very functional, complement the skin tones and are very pleasing when viewed on a person's face.

NISO and QISO photochromic compounds are oxazines that have been described in the literature (Ono and Osada, 1971; Ono *et al.*, 1971; Hovey *et al.*, 1980, 1982; Chu, 1983a; Manfred and Martinuzzi, 1984; Kwak and Hurditch, 1988). These compounds, when incorporated into a lens, typically yield a blue hue when UV-activated, with a maximum absorbance wavelength nominally in the 575–625 nm range. These photochromics are fairly narrow-bandwidth absorbers with half-power bandwidths in the range of 90–110 nm. The activated spectra of two NISO photochromic lenses are shown in Figure 2.9. Such photochromic lenses are not very functional with

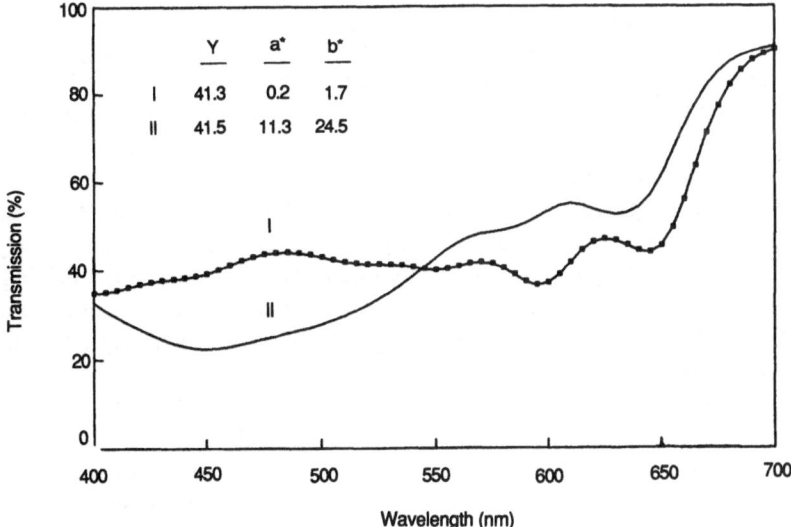

Figure 2.8 Dispersed dyed polymerized CR-39® monomer broad-band spectral filters: (I) gray-tinted lens, (II) brown-tinted lens.

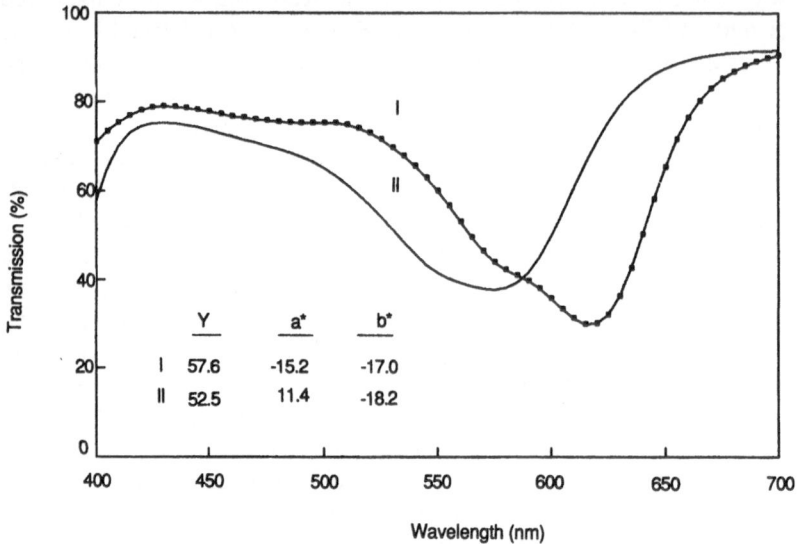

Figure 2.9 Visible spectra of two activated NISO photochromic polymerized CR-39® monomer lenses: (I) 9'-methoxy-1,3,3-trimethylspiro[indoline-2,3'{3H}naphth-{2-1,b}{1,4}oxazine], λ_{max} = 622 nm; (II) 6'-piperidine-9'-methoxy-1,3,3-trimethylspiro[indoline-2,3'{3H} naphth-{2,1-b}{1,4}oxazine], λ_{max} = 578 nm.

respect to visual acuity. A broader band absorbing activated lens can be obtained by tinting the lens with a brown dye of the proper tint and density. The activated hue of the tinted photochromic lens may vary from a greenish-gray, to gray, to grayish-blue, depending upon the initial tint. The broader the activated photochromic absorption band, the less brown tint (higher transmission) is required to attain the desired neutral hue. For an ophthalmic lens, it appears that the bleach transmission should be 80% or greater. If the lens is to be grayish in the activated state, then the activated response of the photochromic must be at best only modest. Lower transmissions may provide inadequate light levels for many indoor functions and night driving, and may be esthetically less acceptable. For sunglass applications, the bleached transmission may be in the range of 40–55%, depending upon the needs of the sunglass-wearer.

Broad-band absorption in a photochromic lens has been described (Casilli, 1989; Crano et al., 1989) where two or more photochromics are intimately mixed together. Such photochromics would have wavelengths of maximum absorbance separated by 75–150 nm. When mixed in the proper ratio, the resulting activated hue is more neutral. The mixing of colors to develop a specific hue is commonly practiced in the dyeing and pigment industries, such as textiles, paints, etc.

Photochromic lenses produced from NISO and QISO photochromics are not the panacea for all types of lighting situations. For instance, they basically are not activated when placed behind the windshield of a car. The windshield is a very good UV absorber, thereby filtering a significant percentage of the UV-A required to activate the photochromic. They also do not operate in situations where the UV is present as a high percentage of scattered or skylight rays. Such conditions exist in the early morning and late afternoon when the sun's rays are traversing progressively more air mass.

2.3.3 Kinetics of photochromic lens processes

There are two rate processes that are of utmost important in order that a photochromic may be considered for a lens application. These rates are those of UV activation and thermal bleaching. Many factors affect these kinetic processes, for example electronics and steric effects of the photochromic; intensity of the UV; temperature; concentration; the plastic matrix into which the specific photochromic is incorporated, etc. Both of the kinetic processes may affect the ultimate density that the activated photochromic lens will attain.

Since the photochromic activation process is proportional to the intensity of the UV, the amount of UV-A from the sun that a lens-wearer will experience is very important. In reality, the output of UV-A from the sun can be quite variable. It depends upon the time of day, time of year, location, angle of exposure, atmospheric conditions, etc. The UV power distribution from the sun progressively increases over the wavelength range 300–390 nm, which represents approximately 5% of the total solar radiant power. Therefore, it is imperative that the photochromics exhibit broad absorption bands (high quantum yields) in the UV region for maximum utilization of the available solar energy. Photochromics with their major absorption bands in the shorter wavelength range will exhibit significantly attenuated response under conditions where the short UV rays are absorbed or scattered, such as, smog, haze, slight overcast, sun low on the horizon, etc. Also, to maximize the activation process, the photochromic must exhibit a low quantum yield for the competitive process of photobleaching.

An acceptable photochromic lens should exhibit some minimal UV activation and thermal bleach rate performance properties. The glass photochromic product (containing silver halides) has been widely accepted by the eyewear consumer, particularly the Photogray Extra® lens, produced by Corning Co. (1988). The activation process for the Photogray Extra® lens, say at 72°F, attains 50% of its ultimate density within 60 s of UV exposure. This means that within approximately 6–7 min the lens will nearly attain its ultimate density. Similarly, at 72°F, the lens thermally bleaches to

50% of its fully bleached transmission state within a 120–140 s interval. This will provide a nearly completely bleached lens within approximately 15–20 min. It must be remembered the 'cones' in the eye require approximately 5–10 min to become fully adapted.

The photochromic lens-wearer expects a certain minimal level of equilibrium state response performance. The Photogray Extra® lens exhibits a greater than 50% change in the equilibrium density of the lens at 95°F. Therefore, a lens with a bleached transmission of 90% would activate to a transmission of less than 45%. All photochromics, including the silver halides and the NISO photochromics, are temperature-dependent in that the activated equilibrium transmission increases as the temperature increases.

2.3.4 Fatigue resistance (eyewear lifetime)

The useful lifetime of the photochromic lens is of utmost importance to its commercial success. Statistics compiled by the eyewear industry report that, in general, an eyeglass-wearer will be refitted with new eyeglasses on an average every 2 years (Ferrara, 1990). Therefore, during this period, the UV-activated equilibrium state response performance of the photochromic should not decrease (loss in response) appreciably. It has been suggested that the relative loss in response be less than 25% over the 2-year interval (Rodenstock US Lens Division, 1989). Response losses that approach 50% are noticeable to the wearer even though the change occurs over a relatively long period of time. The quarter-life performance over a 2-year interval can best be illustrated with respect to a wearer's daily solar exposure. For the average person, the daily, mid-summer, noontime exposure is probably less than 5 hours per week (Urbach et al., 1972). Therefore, the quarter-life of the lens is less than 500 hours of solar exposure. This certainly would not apply to persons with lifestyles or occupations which require extended daily solar exposures.

The effective lifetime of a photochromic lens is dependent upon several factors, such as the specific photochromic used, the amount of photochromic in the lens, the lens matrix material and the utilization of stabilizers.

NISO and QISO photochromics are similar to most other organic molecules in that they photodegrade and/or photo-oxidize when exposed to UV radiation. This photoinduced desensitization process is often referred to as 'fatigue'. Although many factors affect photodegradation, the fatigue of a specific photochromic in a lens is basically a function of the number of UV photons that it absorbs. NISO and QISO photochromics are inherently UV absorbers; therefore, there is a lifetime performance benefit to introduce as many photochromic molecules as possible into a lens, since the outer molecules will screen the inner molecules from photodegradation. However, limitations exist as to the total level of photochromic that can be incorporated. This can depend upon both the photochromic and the lens matrix, in terms of solubility, residual bleach hue, etc. Higher levels of photochromics are of little value with respect to the extension of the wearer lifetime if the photochromic photodegrades to a colored species, particularly if it is a yellowish hue. Generally, photochromic levels in the range of 0.15–0.35 mg/cm^2 of the lens surface are reasonable (Hovey *et al.*, 1980, 1982).

The fatigue resistance of a NISO photochromic lens is reported to be significantly increased by the incorporation of stabilizers such as UV absorbers, hindered amine light stabilizers (HALS) and certain organonickel complexes (Chu, 1983b, 1985b, 1986a, 1988a). Chu (1983c, 1986b) has shown that organonickel complexes can increase the NISO photochromic lifetime expectancy by a factor of 2–3. The utilization of a specific stabilizer depends upon the ability to incorporate it into a specific lens matrix at the required concentration levels.

The extent of fatigue is measured by determining the relative response performance of the photochromic lens under a fixed set of test conditions. The *response test* is best conducted under a solar-simulated xenon lamp illumination which is set at air mass 2 conditions with the temperature of the lens controlled within fairly narrow limits. The fatigue of the lens, generally expressed as a percentage, is determined by measuring the response performance *before* and *after* a specific interval of solar or solar simulation exposure.

The exposure of a lens is best conducted under solar conditions. For actual 'wearer use' comparison, it is best to position the lens in a vertical mode fastened with a 1-inch off-set from the surface of clean marine plywood (Rosenthal *et al.*, 1985). The exposure panel is slowly rotated (~3 rpm) in a 360° arc. This simulates as closely as possible the exposure that a lens would receive on a person's face. This type of comparative testing should be conducted over an extended period of time, preferably starting in the spring and extending into the fall, which gives the basis for attaining good average data. For the fatigue testing of lenses for 'non-wearer' use comparisons, it is best to solar expose the lenses mounted similarly as above but at a 45° exposure angle. This affords a more consistent direct UV exposure for year-round comparative testing than the 90° exposure angle. Under this type of testing, it is best to expose the lenses for a specific total energy. At a South Florida Test Service site near Miami, the average daily energy density at the 45° exposure angle is 0.894 MJ/m^2 (South Florida Test Service, 1989). Therefore, for year-round fatigue-testing comparisons, lenses are exposed for a specific energy interval which is dependent upon both the length of exposure time and the UV intensity.

Accelerated and comparative fatigue testing can also be conducted under solar-simulated conditions. Generally, only the xenon lamp or the UV fluorescent lamp is recommended for solar simulation testing of NISO and

QISO photochromic lenses. The conventional carbon arc lamp is not recommended. The carbon arc contains some very high-energy, short-wavelength UV rays that are not normally contained in solar UV because of the filtration by the earth's atmosphere. The xenon lamp more closely matches sunlight. By the utilization of special filters, the spectral profile can be modeled over a broad range. The profile of a typical xenon lamp operating at two different power densities to simulate both summer and winter sunlight is shown in Figure 2.10. The utilization of a fluorescent lamp for UV exposure is also effective. The UVA-340 lamps (with a wavelength of maximum absorbance at 340 nm, see Figure 2.11; Brennan and Fedor, 1989) simulate very closely parallel sunlight in the wavelength range 295–350 nm and yield, in general, excellent comparative results. The fluorescent lamp–UV solar simulated testing offers both an accelerated and economical mode for testing since many lenses can be UV-exposed simultaneously.

2.4 Structure–property relationships of spiroxazines

The first published reports of photochromic spironaphthoxazines appeared in 1970. There was essentially no activity in this area between the disclosures of Ono and Osada (1970) or Arnold and Vollmer (1970) and the United

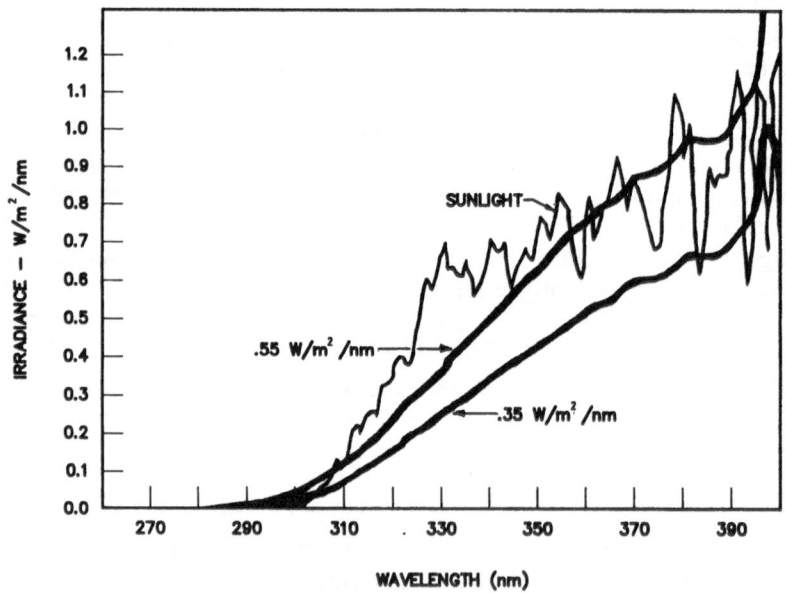

Figure 2.10 Comparison of the power density output of sunlight with that of a xenon lamp as a function of wavelength [0.55 W/m^2 (summer) and 0.35 W/m^2 (winter), at 340 nm]. (Courtesy of Q-Panel Co.).

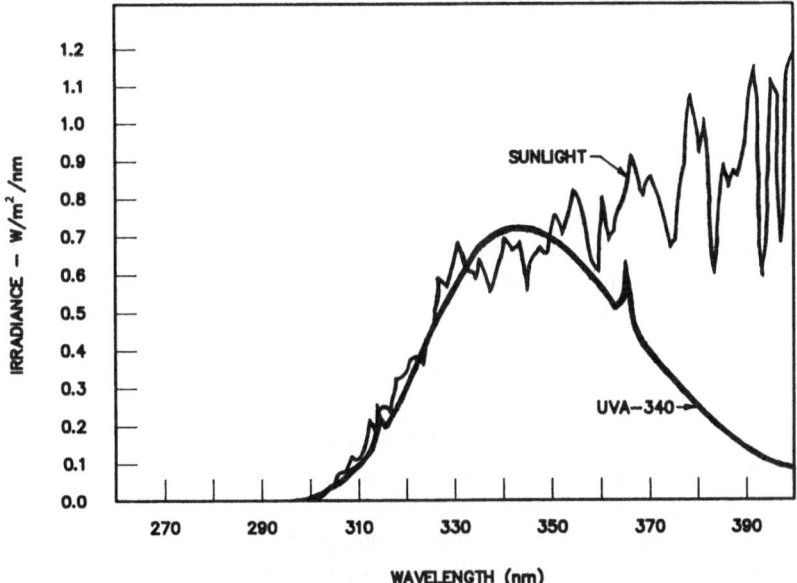

Figure 2.11 Comparison of the power density output of sunlight *vs* that of a fluorescent UVA-340 lamp as a function of wavelength. (Courtesy of the Q-Panel Co.)

States patent issued to Hovey *et al.* (1980). During the 1980s there was a rapid acceleration of research and publications on spiroxazines. The vast majority of the publications available at this time are patents and patent applications. Because of the obscure nature of the properties for specific compounds given in most patents, gleaning the structure–property relationship for this class of photochromic compounds is difficult. For the same reason, some of the data presented in this section are less quantitative than would be desirable. For example, in some cases a shift in a visible absorption band is stated as a color change rather than a quantitative shift in the location of the absorption maximum.

2.4.1 Indolinospiroxazines

The compounds that have received the most attention are the indolinospironaphthoxazines and pyridobenzoxazines. The discussion of these classes of compounds will first be concerned with the indolinospironaphthoxazines (NISO, **1**) derived from unsubstituted and substituted 1-nitroso-2-naphthol and then the pyridobenzoxazines (QISO, **2**) although there will be some overlap of substituent effects.

The purpose of this section is not to review exhaustively all of the literature but rather to indicate apparent trends. Whenever possible, the properties related are those of the compound dissolved in a polymer matrix. However, some data are also presented which have been obtained on solutions in common solvents and where the properties of the polymer solution have not been documented.

2.4.1.1 Indolinospironaphthoxazines (NISO). The substituent effects will be discussed starting with the 1 and 3 positions, followed by a discussion of substituents on the other positions of the indolino ring system. Then, the substituent effects in positions 5′ through 10′ on the naphthoxazine ring system will be examined. In the section on pyridobenzoxazines, some brief comments will be made about the profound effect that occurs when the hydrogen in the 2′ position is replaced with a methyl group.

In liquid or solid solution, the parent indolinospironaphthoxazine (**1**) turns blue upon irradiation with UV light and rapidly fades back to colorless when the activating radiation is removed. Only UV-A light (315–380 nm) is required for activation. It is generally recognized that the blue color is caused by the formation of the ring-opened structure, **3** (MC), which absorbs in the region of 600 nm.

The absorption spectrum of **1** in solution is presented in Figure 2.12. The spectrum shows the weak to moderate absorption bands between 380 and 315 nm which are responsible for the transformation to the colored form under UV-A exposure. The visible spectrum of the colored form in poly(methyl methacrylate) is given in Figure 2.13. The maximum absorption at 612 nm and the shoulder at 578 nm give the shape that is very characteristic for all spironaphthoxazines. Repeating what was illustrated earlier, the very slight effect of solvent on the visible absorption spectrum of the colored form has been documented by Kellmann *et al.* (1989). In methyl

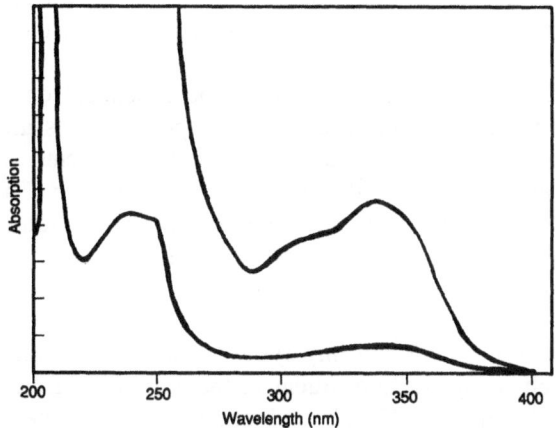

Figure 2.12 Ultraviolet absorption spectrum of NISO (I) in diglyme.

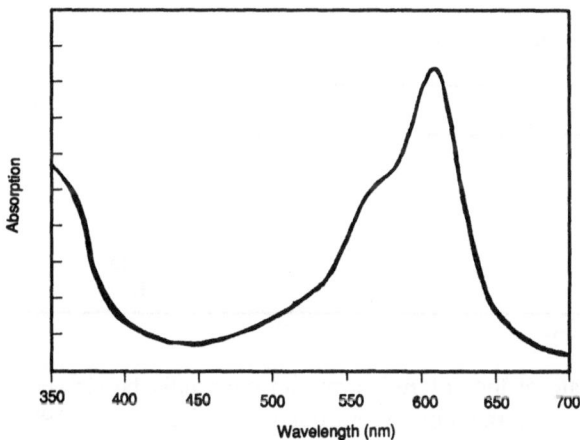

Figure 2.13 Visible absorption spectrum of NISO (I) in poly(methyl methacrylate) after UV activation.

alcohol, the absorption band maximum occurred at 610 nm. Changing to the less polar solvent, toluene, resulted in an hypsochromic shift to about 590 nm. This small degree of solvatochromism is also typical of the substituted spironaphthoxazines.

The thermal fade of **1** has been shown to be dependent on the matrix. The rapid fade in ethanol solution at room temperature was measured by Chu (1983a). In a polymer matrix, fading of the colored species was retarded with the rate inversely related to the stiffness of the polymer (Chu, 1988b). Tateoka *et al.* (1988) reported that the fade rate for the colored form could be increased by the addition of a plasticizer (e.g. dibutyl phthalate) to poly(vinyl butyral).

2.4.1.2 Substitutions on indolino ring system.   Little documentation exists of the effect of changing the substituent in the 1 position from methyl to other alkyl or aralkyl groups. The photochromic response of the spiroxazine was shown to be somewhat affected by the substituent on the indolino nitrogen (Hovey *et al.*, 1982). This is illustrated by the data presented in Table 2.1. For various 1-alkyl-5,6-dimethyl-9'-methoxy NISOs in cellulose acetate butyrate, the photochromic activity increased in the order 1-methyl < 1-ethyl < 1-*n*-propyl < 1-*n*-butyl. However, the effect is not straightforward. For example, the 1-isopropyl-NISO was less active than the 1-*n*-propyl compound and the corresponding 1-*n*-octyl-NISO had less activity than the 1-*n*-propyl or 1-*n*-butyl compounds.

Table 2.1 Photochromic response of 1-alkyl-5,6-(or 4-)-dimethyl-9'-methoxy-NISO in cellulose acetate butyrate.

1-alkyl group	Photochromic response (ΔOD)
Methyl	0.84
Ethyl	0.92
n-Propyl	1.35
Iso-propyl	0.98
n-Butyl	1.45
n-Octyl	1.09

OD, optical density.

The fade rate of the colored form can reportedly be slowed by placing a long alkyl chain on the indolino nitrogen (Yoshitake *et al.*, 1989). Replacing the methyl with benzyl or substituted benzyl was studied by Yamamoto and Taniguchi (1986a, b, 1988). An improvement in fatigue resistance was claimed for benzyl substitution but no data were given to document the effect. The hypsochromic shift in the absorption band upon addition of electronegative groups to the benzyl group on the indolino nitrogen was well documented by relating the color of the activated spiroxazine to the substitution pattern. For example, the activated forms of compounds **4** and **5**

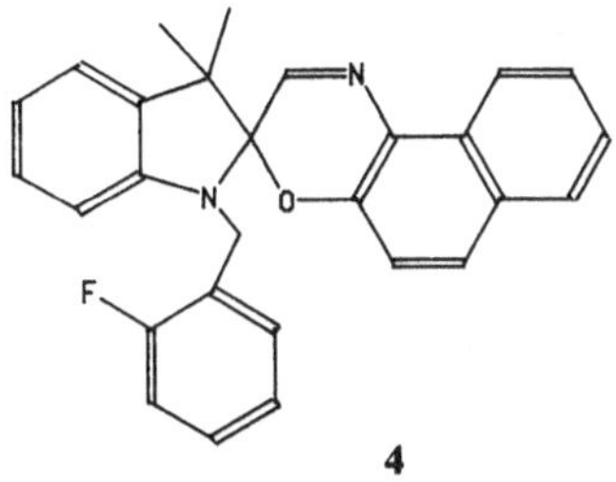

4

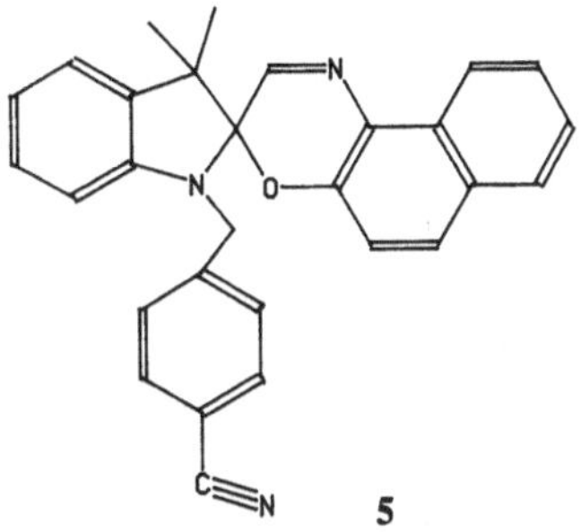

5

with a fluoro- or cyano-substituted benzyl group were reportedly blue–violet rather than blue in color.

Chu (1987) reported that addition of a trifluoromethyl group to the 4 or 6 position of indolinospironaphthoxazine resulted in a similar hypsochromic shift. In a cellulose acetate butyrate matrix, compound 6 exhibited absorption bands at 570 and 590 nm when exposed to UV radiation compared with an absorption maximum at 610 nm and a shoulder at 570 nm for compound 1. Moving the trifluoromethyl group to the 5 position resulted in a 10-nm bathochromic shift, indicating an attenuation of the effect of this electronegative group.

The same general phenomenon was reported by Maltman and Threlfall

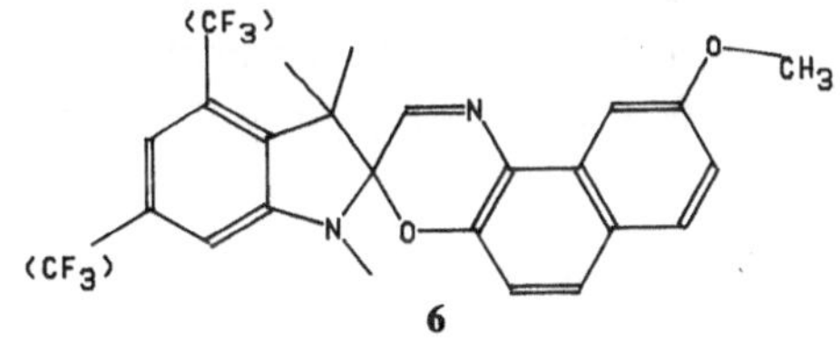

6

(1987) along with the opposite shift in the absorption band with the addition of an electron-donating group, methoxy, in the 5 position. In the cross-linked polymer derived form triethylene dimethacrylate, the open form of unsubstituted indolinospironaphthoxazine had an absorption maximum at 610 nm compared with 620 nm for the 5-methoxy derivative.

The increased photochromic response that results from the addition of electron-donating groups to the 4 through 7 positions of the indolino ring system was documented by Hovey et al. (1980). The placement of methyl or methoxy groups on the 4 through 7 positions of NISO generally improved the activity over that of unsubstituted NISO, but the degree of improvement was dependent on the position substituted (Table 2.2).

Table 2.2 Photochromic response of substituted NISO in cellulose acetate butyrate.

Substituent	Photochromic response (ΔOD)
None	0.9
5-Methyl	1.4
6-Methyl	1.1
7-Methyl	0.8
4(or 6), 5-Dimethyl	1.1
4,7-Dimethyl	1.3
5-Methoxy	2.7
4,7-Dimethoxy	1.2

OD, optical density.

Essentially no results have been published on the effects of varying the substituents in the 3 position. Melzig (1987) claimed that the temperature dependence of the photochromic response could be reduced by replacing the two methyl groups in this position with ethyl groups. However, the data presented indicate this effect to be marginal at best.

2.4.1.3 Substitution on the naphthoxazine ring system. Improvements in photochromic activity obtained through substitution on the naphthoxazine ring system are well documented. The earliest work involved demonstrated the increase in response resulting from the addition of a methoxy group in the 9' position or a bromine in the 8' position (Hovey *et al.*, 1980). This is illustrated in Table 2.3 listing the equilibrium photochromic response, or change in optical density, resulting from irradiation of cellulose acetate butyrate films containing NISO, 9'-methoxy-NISO, and 8'-bromo-NISO. Although dramatically improving the photochromic response, these substituents had little effect on the position of the visible absorption band.

Table 2.3 Photochromic response of substituted NISO in cellulose acetate butyrate.

Substituent	Photochromic response (ΔDO)
None	0.9
9'-Methoxy	1.4
9'-Ethoxy	1.4
8'-Bromo	1.2

OD, optical density.

Dateoka and Sagawa (1987) reported the result of adding the strongly electronegative group, trifluoromethyl, to either the 6' or 8' position. Primarily, the effect was an improvement in the photochromic response at higher temperatures (45°C). No information is given on the amount of residual color observed with the non-irradiated samples.

More interesting is the result of attaching an amino moiety onto the 6' position of NISO. Both Rickwood and Hepworth (1987) and Casilli *et al.* (1989) reported that the addition of an alicyclic amino group in the 6' position causes a hypsochromic shift in the visible absorption band of the activated compound. Whereas NISO yields a blue color upon UV irradiation in a poly(methyl methacrylate) matrix, 6'- piperidino-NISO (**7**) became violet or purple upon activation. The other result of this substitution is a decrease in the inherent fatigue resistance of the material. Replacing the alicyclic amino group with an aromatic amino group, e.g. indolino (compound **8**), shifted the color back to blue.

Melzig (1989) has claimed that the addition of an heteroaromatic group, such as oxazolyl or thiazolyl, in the 5', 6', 8', or 9' positions results in a 20–40 mm bathochromic shift for the colored species. The resulting color is changed from blue to blue–green.

7   8

The addition of a nitro group, common in spiropyran compounds, was studied in the naphthoxazine family by Nedoshivin *et al.* (1989). As expected, a nitro group in the 9' position increased the lifetime of the colored species.

The presence of a methoxy group in the 5' position (**9**) caused a shift in the thermal equilibrium between the uncolored and colored species toward the colored species (Chu, 1986b). In cellulose acetate butyrate, this resulted in a lower light transmission at room temperature before UV activation.

9   CH$_3$

Several patent applications have been filed covering the structure and use of indolinospironaphthoxazines containing unsaturated groups in the naphthoxazine ring system (e.g. Akashi and Taniguchi, 1988a; Akashi *et al.*, 1988) or the 1 position of the indolino moiety (e.g. Takuma *et al.*, 1988). The unsaturated groups are usually acryloyl or methacryloyl and are present for the purpose of copolymerizing with other monomers to form polymers with chemically attached photochromic groups. The advantage given for this technique in one case was the immobilization of the photochromic unit, disallowing migration or leaching from the polymer matrix (Akashi and Taniguchi, 1988a). Another interesting application involved the formation of water-soluble photochromic resins through copolymerization of the photochromic with acrylic functionality with monomers such as acrylamide or *N*-vinyl-2-pyrrolidinone (Akashi *et al.*, 1988).

The formation of water-soluble or partially water-soluble photochromic spironaphthoxazines was accomplished by the addition of sodium sulfonate (Hosoda, 1986; Tamaki and Ichimura 1988) or carboxylate groups (Unitika, 1986).

2.4.1.4 Other indolinospiroxazines. Most of the reports of indolinospiroxazines involve members of the spironaphthoxazine family (NISO) derived from substituted or unsubstituted 1-nitroso-2-naphthols. There have also been several references to spiroxazines other than these naphthoxazines.

The simplest modification would be the spironaphthoxazine (**10**) formed from 2-nitroso-1-naphthol. Only one reference has appeared describing this type of naphthoxazine (Yamamoto and Taniguchi, 1987). The simplest compound (**10**) was reported to be blue–purple in poly(methyl methacrylate).

10

Benzannelation of the naphthoxazine moiety has been accomplished to yield compounds **11** (Dateoka and Sagawa, 1987), **12**, (Tanaka and Kida, 1987), and **13** (Tanaka *et al.*, 1987). Compound **11** (R = *n*-propyl) is reported to be photochromic at temperatures above 50°C. The corresponding naphthoxazine was essentially not photochromic above 45°C. The visible absorption band of the anthracene derivative, **12** (R = methyl), when UV-activated was found to have the same wavelength as that of the naphthalene derivative [610 nm in poly(methyl methacrylate)].

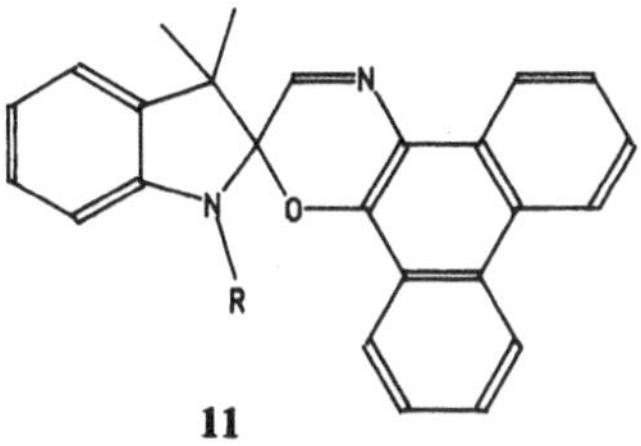

11

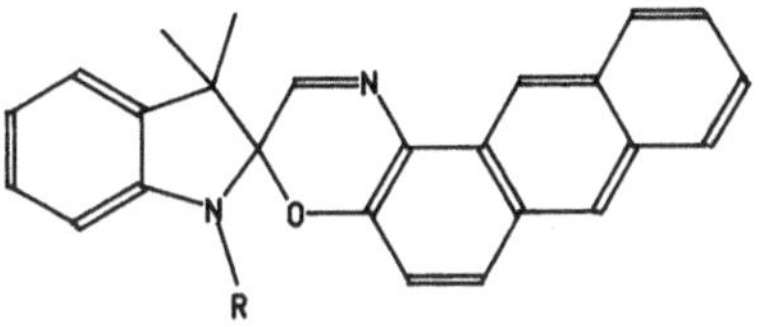

12

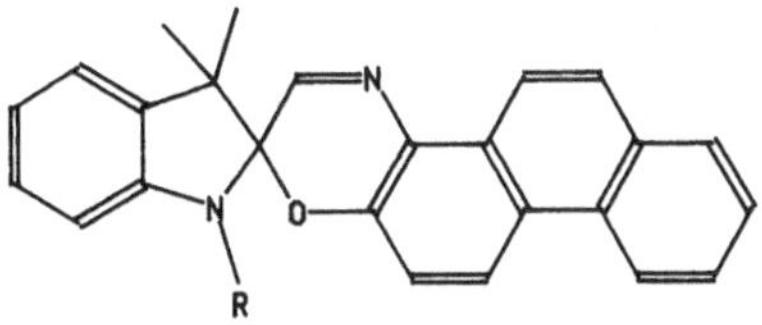

13

The indolinospiropyridobenzoxazines (QISO) based on 6-hydroxy-5-nitrosoquinoline were first reported by Kwak and Hurditch (1987). The insertion of the nitrogen into the naphthoxazine ring system resulted in a large increase in the photochromic response. Melzig (1987) determined that the temperature dependence of the photochromic response of the parent QISO (2) was less than that of 9'-methoxy-NISO.

The pyridobenzoxazines derived from isoquinoline have also been disclosed. Compound 14 (Machida *et al.*, 1990) had a higher level of photochromic response than the corresponding naphthoxazine in a polymer matrix. With similar substitution patterns, the position of the visible absorption band was nearly identical for the isoquinoline, quinoline and naphthalene derivatives.

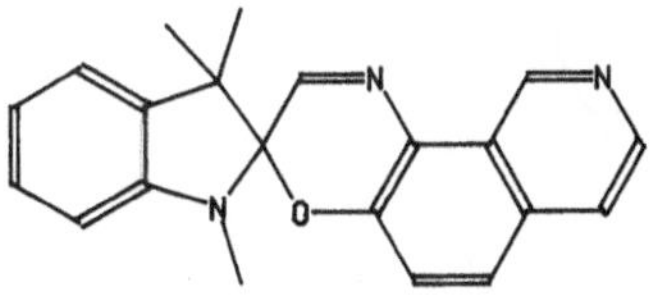

14

Clegg *et al* (1990) have studied the effect of 2'-methyl substitution on the QISO structure (e.g. 15). The addition of the methyl group next to the oxazine nitrogen was found to inhibit completely the photochromic response.

The indolinospiroxazines derived from amino-substituted nitrosophenols were first disclosed by Reichenbacher and Czerney (1986). Reichenbacher did not discuss the photochromic behavior of the aminospirobenzoxazines

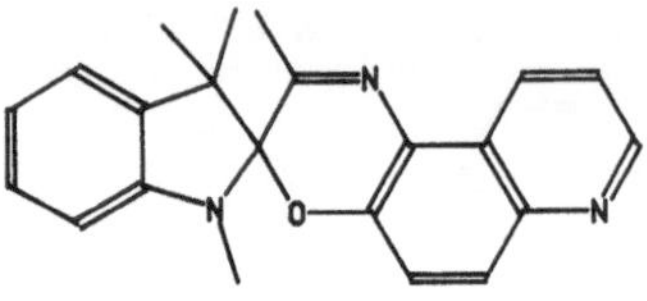

15

but described the reversible formation of stable perchlorate salts. The acid salts (e.g. **16**), which are highly colored, could be neutralized with base to yield the spiroxazine (**17**).

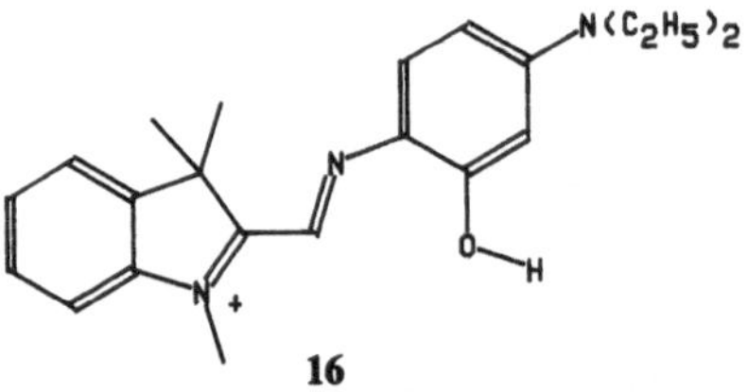

16

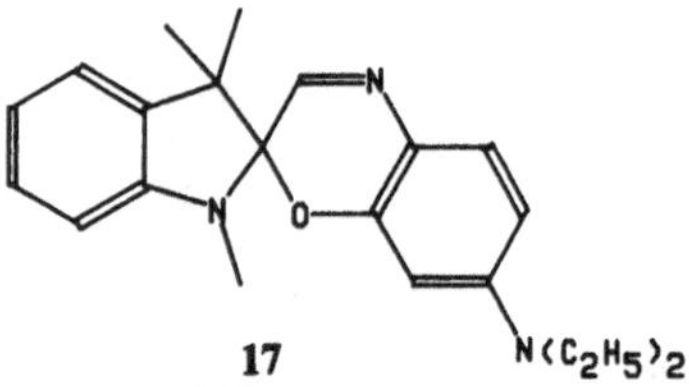

17

The photochromic behaviors of the 7-aminospirobenzoxazines (**18** and **19**) were disclosed by Yamamoto and Taniguchi (1989a, b). The colored species of these were red (λ_{max} = 548 nm or 545 nm in poly(methyl methacrylate)). Substitution on the indolino nitrogen with the group dichlorobenzyl (**20**) caused an hypsochromic shift (λ_{max} = 524 nm).

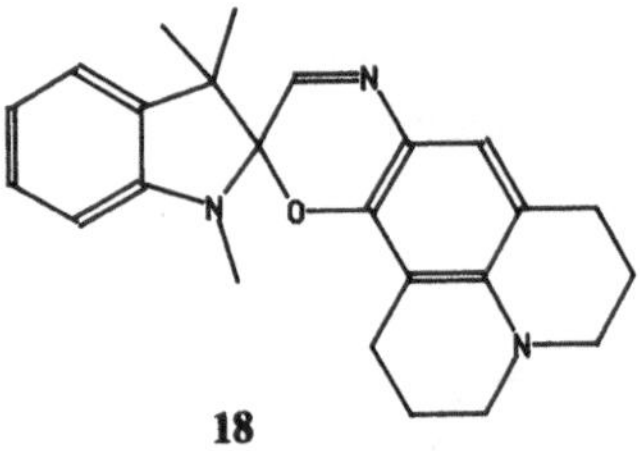

18

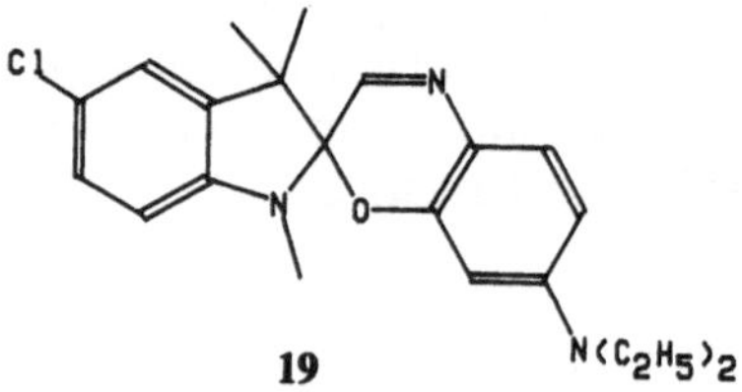

19

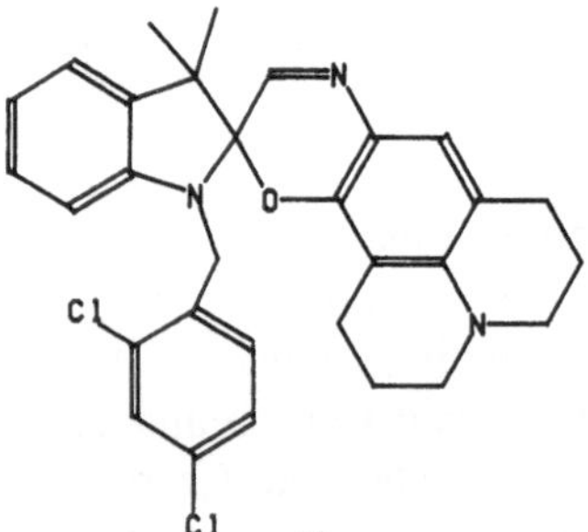

20

Kwak and Chen (1989a) disclosed the photochromic behavior of indolinospirobenzoxazines with a variety of substituents on the benzoxazine ring system. The activated color was reported to be sensitive to the substitution pattern. For example, in ethyl alcohol solution the colored species of compound **21** was orange, whereas that of **22** was blue.

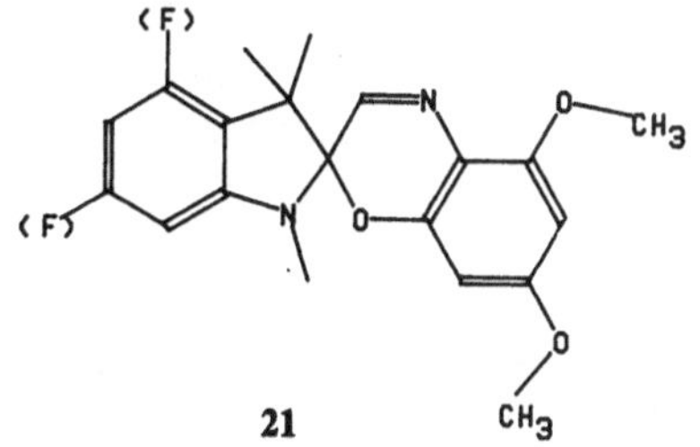

The spiroxazines derived from hydroxynitrosodibenzofurans have been disclosed by Yamamoto and Taniguchi (1989c). These photochromic compounds are interesting because their colored forms have two absorption bands in the visible. For example, compound **23** exhibited absorption bands at 460 nm and 632 nm in methyl alcohol solution after UV irradiation.

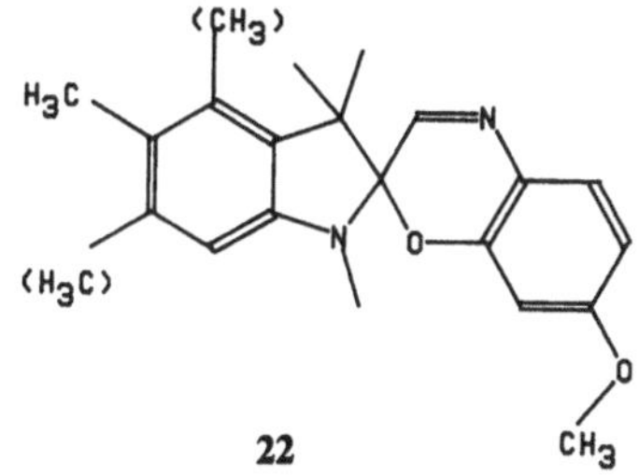

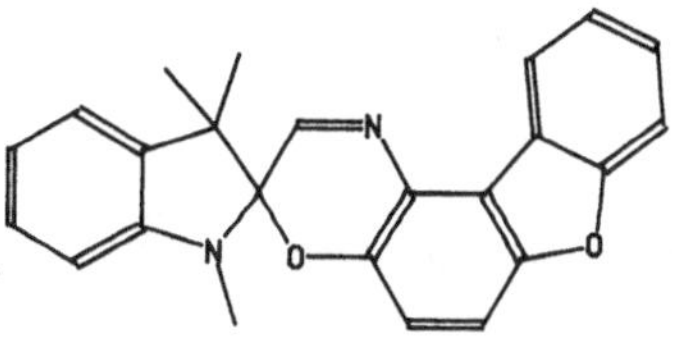

23

Using 1,5-dinitroso-2,6-dihydroxynaphthalene as a starting material, the bis-spiroxazine compound (**24**) was prepared (Kureha Chem. Ind., 1988). Formation of the bis-oxazine structure caused a bathochromic shift in the UV absorption bands of the unactivated compound (λ_{max} = 380 nm) and a similar shift in the absorption band of the colored form [λ_{max} = 630 nm in poly(methyl methacrylate)].

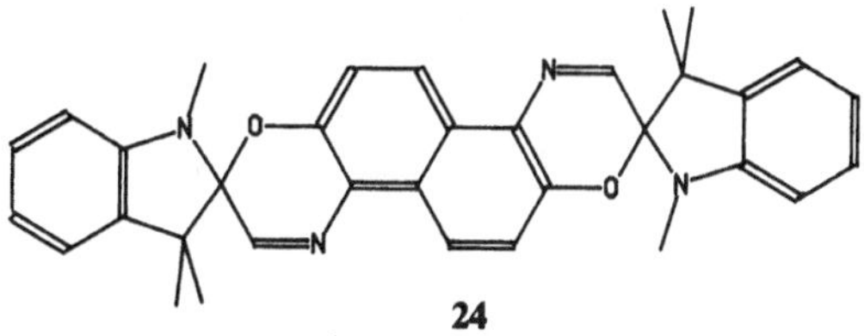

24

2.4.2   Other spiroxazines

Replacing the indolino group of the spiroxazines with piperidino has been reported by Kawauchi *et al.* (1987, 1990), Matsuoka (1988) and Yamamoto and Taniguchi (1989d). As stated before, NISO (**1**) possesses a blue activated form (λ_{max} = 610 nm) in liquid solutions or polymer matrices. In contrast to this, the activated form of the piperidino compound (**25**) was purple–magenta (λ_{max} = 580 nm) (Kawauchi *et al.*, 1987) or purplish-red (Matsuoka, 1987) in toluene. In methyl alcohol, the visible absorption band maximum was at 562 nm (Kawauchi *et al.*, 1990). The fatigue resistance of **25** was comparable to that of NISO.

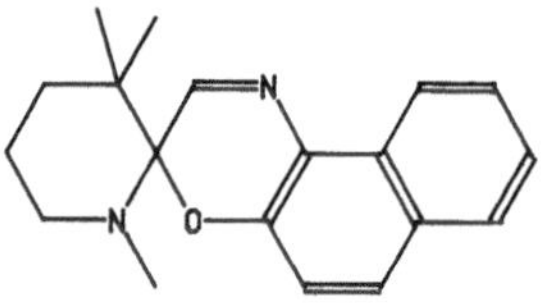

25

The piperidinospiroxazine (**26**) derived from 2-nitroso-1-naphthol was prepared by Yamamoto and Taniguchi (1987). The activated form was violet in poly(vinyl butyral).

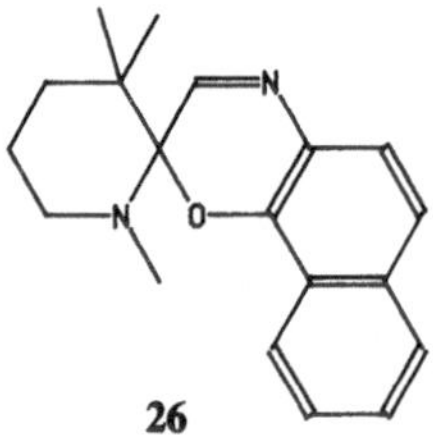

26

Kwiatkowski and Hunt (1988) have disclosed the oxazolidinospiroxazines, such as **27**. This compound formed a red species upon UV irradiation in toluene which reverted to the uncolored **27** after removal of the UV radiation. This photochromic behavior is in contrast to that of indolinospiroxazines,

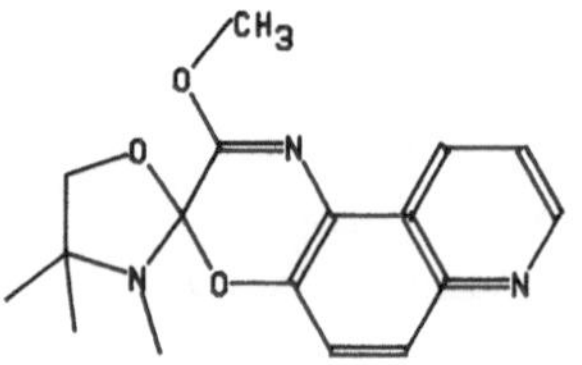

27

which when substituted on the oxazine ring inhibited a photochromic response (Clegg *et al.*, 1990).

Yamamoto and Taniguchi (1989e) have disclosed the photochromic behavior of the benzothiazole derivative **28**. The colored form was reportedly purple in poly(methyl methacrylate).

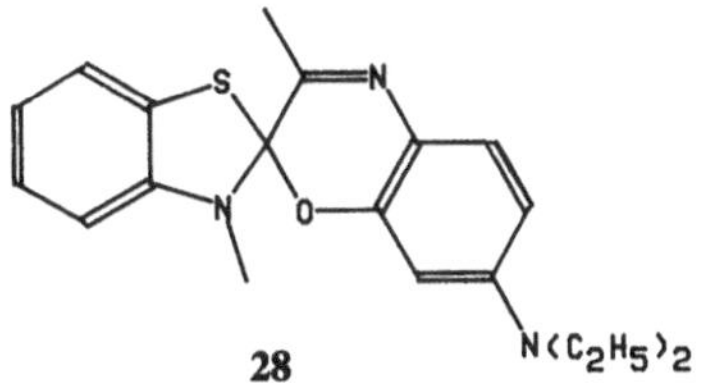

28

Melzig (1990) has reported that the adamantyl spiroxazines such as **29** are photochromic but no details are disclosed.

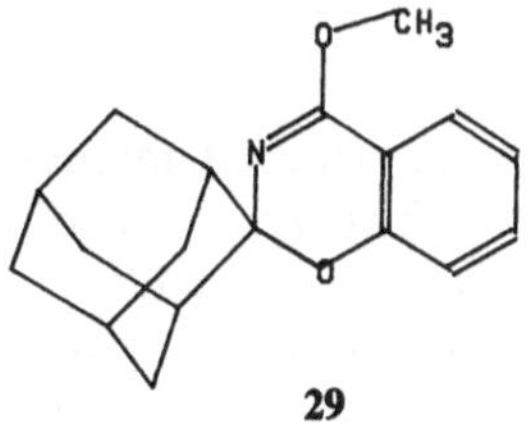

29

2.5 Stabilization of spiroxazines

One of the most important properties of any photochromic compound, if it is to be used in a plastic lens system, is stability against photodegradation. Good photostability, also called fatigue resistance, is required for a practical lifetime for the lens system.

Indolinospironaphthoxazines are inherently more fatigue-resistant than the spiropyrans. As measured by the quantum yield for photodegradation, the spironaphthoxazines are two to three orders of magnitude more photostable than the spirobenzopyrans (Chu, 1984a). Nevertheless, spiroxazines are reported to be stabilized even further by various protective procedures.

The partial stabilization of spiroxazines against photodegradation by isolating the photochromic-containing material from oxygen with an inorganic barrier has been reported (Uhlmann *et al.*, 1979, 1983; Seiko Epson, 1987; Yamamoto and Taniguchi, 1989a). The protection from photo-oxidation was also indicated by Tateoka *et al.* (1987) by showing that the fatigue resistance of the dimeric oxazines, e.g. **30**, was directly related to the thickness of the plastic film containing the photochromic compound.

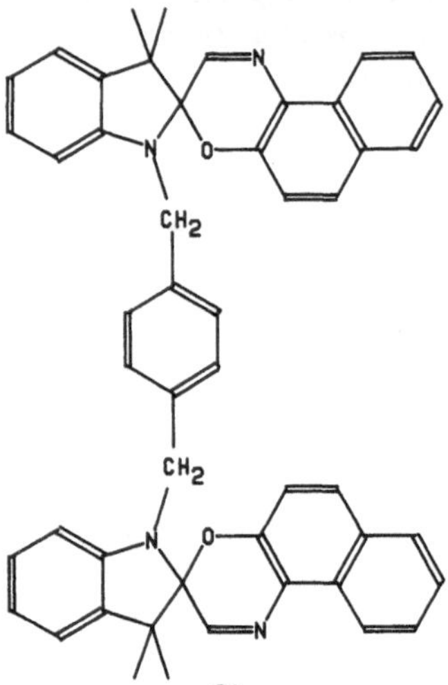

30

Chu (1986b) demonstrated the effect of mixing standard UV stabilizers with NISO in a polymer matrix. This type of additive, which absorbs some UV radiation, slightly enhanced the photostability of NISO but also reduced the magnitude of the photochromic response. An example of a standard UV stabilizer is the hydroxyphenylbenzotriazole, Tinuvin® P (Ciba-Geigy).

Other stabilizers have been used with some success. For example, the addition of the organonickel complex, Cyasorb® 1084 (**31**, American Cyanamid), to cellulose acetate butyrate containing the spironaphthoxazine **5** increased the photostability of the photochromic compound considerably (Chu, 1984b). After 60 h in a UV exposure chamber, the unprotected system had lost all of its activity. In the presence of the nickel complex, 87% of the original activity remained. Other nickel complexes worked as well, for example nickel dithiocarbamate.

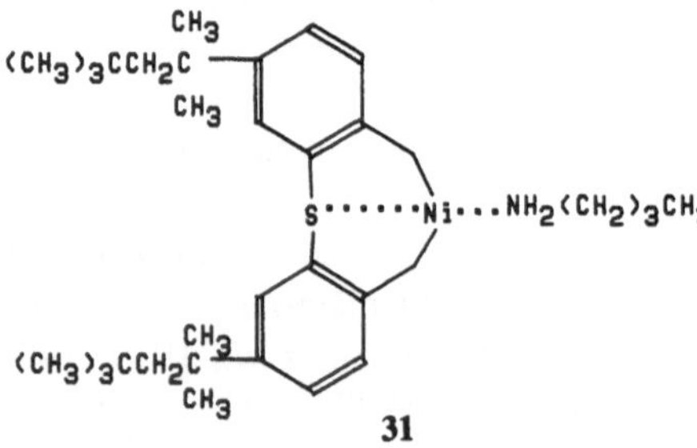

31

Hindered amine light stabilizers (HALS) constitute a class of stabilizers which are effective in protecting spironaphthoxazines without reducing the photochromic activity. Chu (1988b) demonstrated that the addition of a HALS, e.g. Tinuvin 770 (**32**, Ciba-Geigy), to cellulose acetate butyrate films containing the spironaphthoxazine **33** improved the fatigue resistance. After 100 h of UV exposure, the samples without HALS had no remaining photochromic activity. In the presence of Tinuvin 770, only 36% of the original activity was lost after the same exposure period.

The addition of a HALS to a plastic sample containing the photochromic **33** plus a nickel complex significantly enhanced the stability of the system over that obtained with either the nickel complex or HALS alone (Chu, 1988c).

Other stabilizers have been reported to be effective in improving the photostability of spironaphthoxazines. Tateoka *et al.* (1989a) disclosed the stabilization of NISO dissolved in poly(vinyl butyral) with the addition of 9-cyanoanthracene. The use of a nitroxyl free radical as a stabilizer for spiroxazines was reported by Tateoka *et al.* (1989b).

2.6 Development of a commercial system for plastic photochromic eyewear

2.6.1 Plastic lens manufacturing

The lens industry was born and raised with the glass lens. However, in the prescription eyewear industry, the switch to plastic lenses was initially slow but then greatly accelerated with the advent of larger, more fashionable lenses. The basic key is the reduced weight of plastic lenses versus glass, because the density of plastic is only 52% that of glass. The light weight is also a must for persons with relatively high positive- and negative-diopter corrections. Presently, in the US market, over 70% of all prescription eyewear lenses are now fabricated from plastic. The portion of the US market served by glass is basically controlled by the glass (silver halide) photochromic products (~50% of all glass lenses), and glass lenses worn by patients whose occupation may subject the lenses to adverse conditions.

Optical plastics are not only used for eyewear but are widely employed for applications such as glazing, lighting and decorative materials. The unique combination of high light transmission, light weight, dimensional stability and formability, and their chemical, weather and impact resistance contribute to their versatility.

There are three major optical plastics employed in the transparency and eyewear industry. They are polymerized CR-39® monomer (or allyl diglycol carbonate), bisphenol-A polycarbonate and poly(methyl methacrylate). Of these, the first is a thermoset resin, while the last two are thermoplastics. A comparative analysis of some selected important properties of these polymers is shown in Table 2.4.

Table 2.4 Selected properties of some optical plastics*.

Property	ASTM method	Polymerized CR-39® monomer	Poly(methyl methacrylate)	Bisphenol-A polycarbonate
Refractive index (N_D) (589.3 nm)	D542	1.498	1.491	1.586
Abbe number (V)	D542	59.3	57.2	34.0
Haze (%)	D1003	< 1	< 2	< 3
Luminous transmittance (%) (0.125 inches thick)	D1003	92	92	85–91
Deflection temperature (°F) (3.6°F/min, 264 psi)	D648–56	140	198	280
Density (g/cm³)	D792	1.31	1.19	1.20
Rockwell hardness (0.25 inches sample)	D785–62	M95	M97	M70
Impact strength (ft. lb/in) (Izod notch)	D256	0.2–0.4	0.3–0.5	12–17
Thermal expansion (linear coefficient/°F × 10^5)	D696–44	4.7	3.6	3.8
Surfacing ease (relative)		High	Low	Low
Abrasion resistance (relative)		High	Medium	Low
Disperse dyeability (relative)		High	Low	Low
Chemical resistance (relative)		High	Low	Medium

* Data taken from manufacturer's published literature.

Poly(methyl methacrylate) is the most widely used optical plastic; however, it is seldom used as an optical eyewear lens material. It has relatively low impact resistance, poor chemical- and abrasion-resistant properties and, in addition, the lens blanks are difficult to surface and polish to the required prescription and then edge for framing. Poly(methyl methacrylate) is a relatively inexpensive plastic and lenses can be cast individually, either directly from monomer or injection-molded from fused polymer.

Bisphenol-A polycarbonate is a low-density, high index of refraction polymer with excellent impact resistance. It has been widely accepted as the material to be utilized for safety-type lenses. On the negative side, it has only medium chemical resistance and relatively poor abrasion-resistant properties. The slow acceptance of polycarbonate by the prescription eyewear market (presently at the 3–4% level) has been because of its poor surfacing–polishing properties (because of its high ductility), and some optical aberration problems. Another major reason has been the limited availability of a broad range of lens blanks for filling normal prescriptions. The production of each base curve, whether for single vision, multifocal or progressive-plus lenses, requires a specially tooled injection mold.

CR-39® monomer, allyl diglycol carbonate, was first patented by Muskat and Strain (1945) and is still produced by PPG Industries. Polymerized CR-39® monomer is a thermoset resin, generally catalyzed with percarbonates, such as di-isopropylperoxydicarbonate. The catalyzed monomer, along with any additives, is dispensed into a matched pair of precisely machined and highly polished glass or metal molds, separated by a gasket (Figure 2.14). Polymerization occurs under well-controlled extended thermal cycles in an air-recirculated oven or a water bath. Polymerized CR-39® monomer is characterized by its high optical quality and attractive mechanical, thermal, electrical and chemical-resistant properties. The high transmission of greater

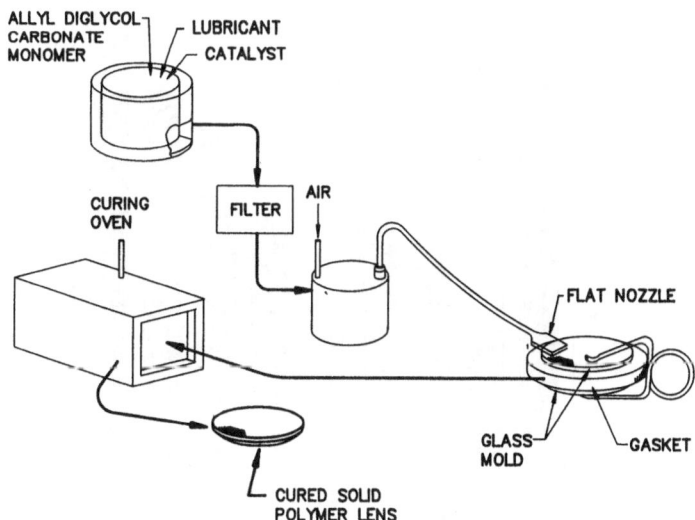

Figure 2.14 Schematic diagram of the casting process for polymerized CR-39® monomer lenses. (Adapted and redrawn from Beattie, 1951.)

than 90% through the wavelength range of 350–1100 nm accounts for its desirable optical properties. Its ease of mechanical working by surfacing, polishing and edging has made it the choice plastic by optical laboratories for fabricating prescription lenses. Another plus is that it can be readily and easily tinted at the optical laboratories or at the dispenser site by simple aqueous disperse dye techniques to a variety of hues and densities. Polymerized CR-39® monomer is consumed almost exclusively in the eyewear industry, and over 90% of all plastic prescription lenses are fabricated from allyl diglycol carbonate. Lenses are cast and shipped around the world.

More recently, there is a trend toward the greater utilization of higher index of refraction (1.55–1.65 range) polymers (perhaps 7% of the present total plastic market; Ferrara, 1990). These polymers generally contain high concentrations of aromatics and may contain halogens. The higher index resins afford patients with either high negative or positive corrections an option for a thinner lens which results in an overall reduced weight than say, with a comparable polymerized CR-39® monomer lens. For example, a negative 8-diopter prescription lens correction fabricated from a 1.6 index plastic will be 20% thinner and 15% lighter than a comparable prescription lens made from a polymerized CR-39® monomer (Figure 2.15).

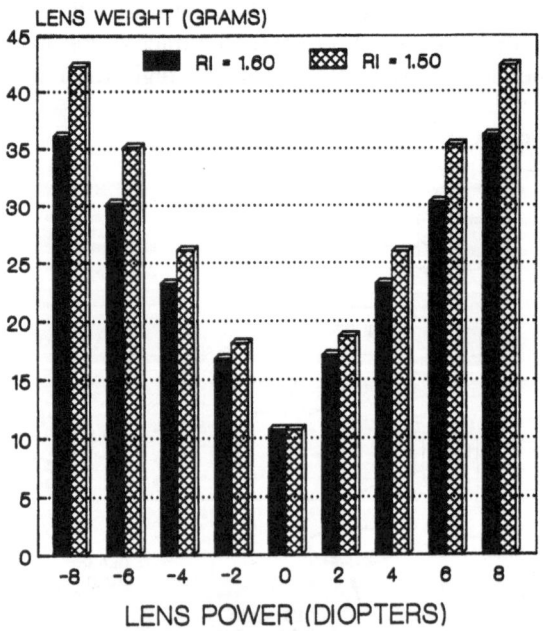

Figure 2.15 The comparison of lens weight for two plastic lenses with refractive indices of 1.50 and 1.60 as a function of lens correction. Based on lenses 72 mm diameter and polymer density 1.31 g/cm³.

Basically, prescription eyewear lenses are manufactured, processed and dispensed as monolithic components. However, several lens concepts have been tested where two lens components are laminated together with either an adhesive or other bonding media (Tolar, 1974). Such concepts attempt to take advantage of the best properties of both glass and plastic by laminating an outer thin glass lens to a thicker plastic lens. Such a lens provides the excellent abrasion resistance of glass with the light weight and tintability of plastic. In some cases, the glass lens can be photochromic. Essilor of France has offered a glass–plastic laminated lens identified as Essilor CS™. Several techniques to fabricate laminated NISO photochromic lenses have been described (Mitsubishi Gas Chem, 1986; Sakagami *et al.*, 1986; Nissan Motor and Mitsubishi Kasei, 1990a). The laminated lens concept has been used to great advantage with the fabrication of polarized lenses where the polarization film is sandwiched between two plastic lenses. Laminated lenses, in general, have not been well received in light of the fact that some of the earlier lenses exhibited a high rate of delamination when thermally cycled. In addition, the type of frame matched with the lens may be somewhat limited because of the required edging on the lens which may apply undue stress to the lamination layers. The application of lamination concepts to fabricate polarized lenses containing NISO photochromics has been described (Toray Industries, 1989a, Taniguchi *et al.*, 1989).

2.6.2 Processes for photochromic incorporation

2.6.2.1 Coatings. The NISO photochromics are significantly more stable against photodegradation than other organic photochromics, as discussed above. However, their relatively modest thermal stability and sensitivity to oxidants control the manner in which they can be incorporated into a lens. Their good photoresponse and relatively high solubility in a broad range of resins enables them to be incorporated into a lens utilizing a number of techniques.

The application of an optical film containing a photochromic to the lens surface has the potential to turn any lens, regardless of composition or shape, into a photochromic lens. Of greatest potential is to apply the photochromic as part of the abrasion-resistant coating which is commonly applied to plastic lenses (Nakajima *et al.*, 1988a, Sakagami *et al.*, 1988). This type of coating is (or can be) applied at either the dispenser or the optical laboratory level, as well by the manufacturer. A preferred method is to dissolve the photochromic into an organic resin (and perhaps solvent) such as an acrylic resin, and then apply this solution to the lens surface by any one of several techniques such as dip, spray, spin, flow coat, etc. (Taniguchi *et al.*, 1989, Ono and Osada, 1971). A water-soluble NISO photochromic has been described which allows organic films to be applied to an acrylate lens surface without organic solvent attack (Akashi and Taniguchi, 1988b).

Another process for applying an optical film containing a photochromic to a lens surface is by 'in-mold coating'. Here, the photochromic/resin film is applied directly to the mold surface onto which the lens is cast either from catalyzed monomer or by the direct injection of polymer. This process has significant cost advantages in that the lens is available to be marketed directly from the casting mold (Toray Industries, 1985).

A lens with a photochromic coating has an inherent limitation with respect to its potential performance. From an adhesion standpoint, the limitation on the thickness of an optical film is approximately 5 μm. On an average, a 5-μm-thick film will contain only approximately 0.6 mg/cm² of resin. However, the amount of a typical NISO photochromic necessary to produce a photochromic lens which will afford good response with a reasonable wearer lifetime performance (fatigue resistance) is from 0.15 to 0.35 mg/cm² (Hovey *et al.*, 1980; Sakagami *et al.*, 1988). This means that the photochromic is approximately 25–50% of the film composition. Depending upon the photochromic, this concentration may require a solubility that is not feasible. In addition, further problems occur when photochromic stabilizers are introduced to decrease the fatigue rate. In many cases, the stabilizer concentration is equal to or exceeds that of the photochromic (Chu, 1988a). In reality, the physical and chemical properties of the film are no longer those of the resin but are significantly altered as a result of dilution by the photochromic and stabilizers.

2.6.2.2 Injection-molded thermoplastics. Indolinospiroxazine photochromics are soluble to some extent in most thermoplastic resins, such as bisphenol-A-type polycarbonates and acrylics (Hovey *et al.*, 1980; Chu, 1986b, 1989; Nissan Motor and Mitsubishi Kasei, 1989). Therefore, it would appear reasonable that injection molding would be an excellent process to produce photochromic lens blanks on a large, cost-effective, commercial scale. Generally, a concentration of 0.1–0.3% of photochromic in the final lens blank should be adequate for both good response and wearer lifetime performance. Most oxazines are temperature-sensitive on prolonged heating above 160°C; however, temperatures as high as 200°C can be tolerated for a few minutes. Therefore, residence time in the extruder becomes critical. Bisphenol-A-type polycarbonates and acrylics are good candidate resins for lens fabrication. American Optical has used injection molding of cellulose acetate butyrate with NISO photochromics to produce a 1.5-mm-thick goggle-type sunglass lens (Chu, 1986b). Injection molding is essentially a single-step process for photochromic lens fabrication.

2.6.2.3 Internally cast with catalyzed monomers. Dissolution of the indolinospiroxazine (at levels of 0.1–0.3%) directly in the catalyzed monomer would be expected to be the most cost-effective and facile process for the production of photochromic lenses. Such a lens could be cast in highly polished and readily available glass or metal single-lens molds. This would also represent, essentially, a single-step process for photochromic lens production. However, oxazines are very sensitive to strong oxidants and free radical initiators, and therefore they are decomposed to a great extent during the curing process. This is especially true of the thermoset polymer of CR-39® monomer, since high levels (3.0–3.5 ppm) of peroxide catalyst are required for a fully cured casting. Consequently, such a process has never been commercialized. A body of literature describes a technique for protecting the photochromic in a core–shell concept where the core, being the photochromic, is surrounded with an alcohol, ester or resin which has an index of refraction equivalent to the cured resin (Canon, 1986; Kido, 1988; Matsui Chem. Ind., 1990).

NISO photochromics have been successfully cast directly in acrylics. With acrylate and diacrylate monomers, diazo catalysts at levels of approximately 0.2% will produce a fully cured photochromic casting with very little decomposition of the NISO photochromic. The specific acrylic utilized can exert a significant influence on both the photochromic UV activation response and thermal bleach performance (Maltman and Threlfall, 1987; Nakajima et al., 1988a, b; Taniguchi, 1988; Yamamoto and Taniguchi, 1989a). Recent publications suggest appending polymerizable functionality onto the NISO photochromic, which would allow it to become part of the polymer backbone when copolymerized with a monomer such as 2-hydroxyethyl methacrylate (Toray Industries, 1990).

2.6.2.4 Lens surface treatment.

Incorporation of a NISO or a QISO photochromic into the resin lens surface can be conducted by one of several processes, such as solvent dyeing, aqueous disperse dyeing and vapor–liquid transfer. In all of these processes, permeation of the photochromic into the lens is characterized by dissolution of the photochromic at the lens surface, followed by its diffusion further into the lens subsurface under a concentration gradient.

Solvent dyeing is exemplified by the immersion of the lens into a fairly high boiling point organic solvent bath containing the dissolved NISO (Hovey et al., 1982; Nakajima et al., 1988b; Casilli, 1988). The solvent is usually an alcohol, glycol, aromatic or an alkane, and the bath is maintained in the range of 90–120°C with a photochromic concentration of 5–10%. The advantage of solvent dyeing is the ease of attaining photochromic incorporation. The disadvantage is that, generally, the solvent will also solubilize and diffuse into the lens surface with the photochromic and, as a result, expand (swell) the lens substrate. If the solvent is not removed slowly under controlled conditions, the lens substrate will irreversibily craze. As a result of the excessive solvent codissolution, high levels of photochromic are difficult to obtain. Thermal stability of the NISO photochromic in the bath is also a major problem, because of its required high concentration and the relatively long dyeing times. Solvent dyeing is generally not applicable to acrylates and bisphenol-A polycarbonate because of the poor chemical stability of these resins. Solvent dyeing is effective for the incorporation of NISO photochromics into polymerized CR-39 monomer. However, owing to the high cost of NISO photochromics, solvent dyeing may not be a commercially viable process for incorporating photochromics into lenses. Hovey (1981) reports producing a gradient NISO photochromic lens by the gradual immersion of the fabricated photochromic lens into a solvent dye bath containing a UV absorber.

Disperse dyeing is similar to solvent dyeing except the NISO photochromic is only very slightly soluble in the dispersing media and the media does not solubilize into the lens surface. The photochromic can be maintained at low levels, typically at less than 1%, and thermal decomposition is significantly less of a problem. Dispersing media may be water, glycerols, mineral oil, silicone oil, or any other relatively pure, low-volatility and non-reactive material (Casilli, 1989). Disperse dyeing would be available for the incorporation of NISO photochromics into all types of lens materials.

An effective and efficient process for incorporating a NISO or QISO photochromic into the surface of the lens is reported to be by vapor–liquid phase transfer (Le Naour-Sene, 1981). The lens can be heated *in vacuo* or under an inert gas atmosphere in the presence of vapors of photochromic where the photochromic is supported away from the lens surface. High loadings of photochromic incorporated into the lens can be effectively attained. An alternative process is to support the photochromic either as a neat solid or liquid in a cellulose layer supported against the surface of the lens. In both of the above processes, the photochromic is transported to the surface of the lens by either a sublimation or vaporization mode where it condenses. The photochromic permeates into the interior of the lens by a dissolution–diffusion mechanism.

In an another process, the NISO or QISO photochromic (loading levels 25–75% of total solids) and a resin are both dissolved in a solvent. The film-forming solution is applied to the surface of the lens by any one of several techniques, such as dipping, spraying, spinning, flow coating, etc. The solvent is flashed off and the coated lens is heated to a sufficient temperature for a sufficient length of time to effect diffusion of the photochromic from the resin film to the lens substrate under a concentration gradient. Generally, the heating temperature is in the range of 135–150°C for several minutes (Welch, 1989); however, temperatures of near 200°C can be used for a few seconds (Le Naour-Sene, 1981). Elevated pressures can reduce the heating time (Melzig, 1985). The higher temperature process may entail the condensation of the liquid photochromic at the resin film–lens surface interface. After the heat treatment step, the desired photochromic lens is reclaimed by removal of the spent photochromic–resin film. Resin films such as poly(methyl methacrylate), poly(vinyl acetate), poly(vinyl chloride), ethyl acetate, poly(vinyl butyrate), etc. are good photochromic film formers. Efficiencies of the photochromic transfer process, from the resin film into the lens can be of the order of 10–50%, depending upon both the specific photochromic and the resin. Very high-quality and high-performing photochromic lenses can be produced by the vapor–liquid transfer process.

2.6.3 The finished lens

In this section, the type of photochromic prescription lenses that are both required and supplied to the eyewear industry are discussed.

2.6.3.1 Human eye defects.
The refractive index of various parts of the human eye varies from 1.33 to 1.40. The total refractive power of the eye is approximately 59 diopters when accommodated for distant objects. This means that the effective focal length for the eye is approximately 17 mm. The diopter is a unit of measurement in optical lens power which may designate curvature, lens focal length or prism power. Lens power is expressed as the reciprocal of focal length when the latter is measured in meters. A lens of 1 diopter of refractive power will bring parallel rays of incident light to a focus at a 1 m distance. When the incoming light rays are bent by the lens to the proper extent, then the image will focus exactly onto the retina; this is normal vision and is called emmetropia. When the image is focused in front of the retina, this is called myopia or near-sightedness, while the case where the image is focused behind the retina is called hyperopia or far-sightedness (Figure 2.16). Single-vision lenses are used to correct the above two vision defects. As a person nears 40–50 years of age, the eye (lens) slowly becomes less elastic and loses its ability to focus quickly from a point at a distance to one that is relatively close. The eye also slowly loses the ability to focus at a comfortable near distance. This whole process is associated with the vision deficiency commonly associated with loss of 'accommodation'. This condition of the eye is called presbyopia. To rectify this defect, multifocal eyewear such as bifocals or trifocals is prescribed.

2.6.3.2 Lens base curve.
Prescription plastics eyewear lenses are manufactured with a specific spherical base curvature on the convex surface. Generally, they are cast in a semifinished state and an individual patient's prescription is then surfaced into the concave (backside) surface of the lens. The semifinished lens may be up to 80 mm in diameter with a thickness of 15 mm. This allows a specific semifinished base-curvature lens to be applied to a fairly broad range of corrections. Generally, five different base

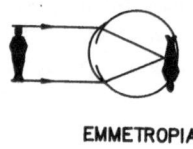

EMMETROPIA

MYOPIA

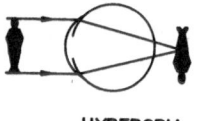

HYPEROPIA

Figure 2.16 Common types of eye defects corrected by prescription eyewear.

curvatures will satisfy the bulk of the corrections normally encountered. Although there is no industry standard, with each manufacturer having slightly different specifications, a general selection of 'base curves' would be 1.0, 2.5, 4.5, 6.25 and 8.25. A prescription for a highly negative-diopter refractive power lens could be fulfilled by surfacing a lower base curved lens blank, while one for a highly positive-diopter refractive power lens would be surfaced from a higher base-curved lens blank. Examples of semifinished lenses and the resulting finished lenses for the above two examples are shown in Figure 2.17. A near-sighted person will require a negative-diopter power

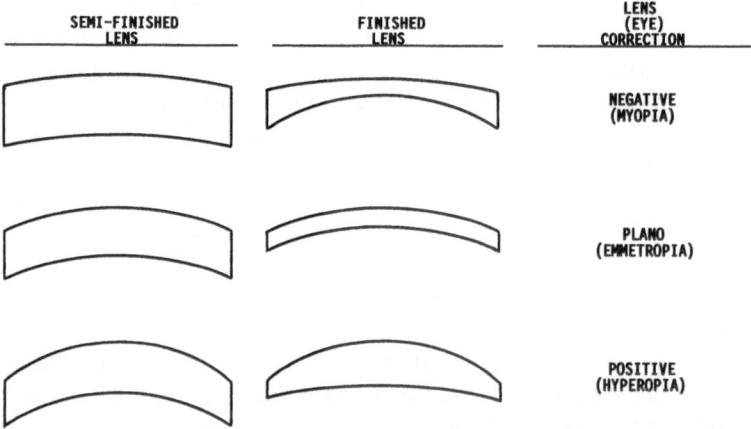

Figure 2.17 Examples of miniscus lenses, in both semifinished and finished form, commonly used for prescription eyewear lenses.

lens for corrected vision, while a far-sighted person will require a positive-diopter power lens. A plano lens has no power.

In plastic lens manufacturing, the added power that is required for both the bifocal and trifocal segment is cast directly into the convex spherical surface. Generally, lenses are produced with the added power segment incremented in 0.25 diopters, up to 3.0 diopters. As one can readily calculate, with the five base curves, left and right eye considered, and 12 combinations of added segment powers, the minimum number of lenses that must be stocked becomes quite large when multifocals are considered. In the US market, 44% of all lenses dispensed are multifocal with the remainder being single vision.

The newest lens on the market with a major impact is the progressive plus (with perhaps 20% of the multifocal market; Ferrara, 1990). In this lens, there is not a sharp break in the lens power correction in progressing from the single vision part of the lens to the higher power region of the lens. Esthetically, these lenses have a very high appeal. However, visually with the power changing gradually across the lens, some blurriness in the peripheral regions can be an annoyance for some patients.

All of the above types of lenses can be incorporated with NISO and QISO photochromics. Therefore, plastic photochromics lenses can serve basically all major aspects of the prescription eyewear market.

2.6.3.3 Commercial plastic photochromic lenses. Presently, 75 million prescription pairs of lenses are dispensed in the US annually. Of these, approximately 10 million pairs (13%) are glass photochromic lenses, produced nearly exclusively by Corning (Optical Manufacturers Association, 1990).

NISO- and QISO-type photochromics, as a class in general, offer UV activation properties and thermal bleach properties which are convenient for the eyewear application. In addition to variable light attenuation, a plastic photochromic lens can offer a number of visual and esthetic enhancement properties, such as: (1) ultraviolet energy protection, both UV-A and UV-B; (2) comfortably light weight; (3) tints from a fashionably light tint to a functionally darker tinted sunlens; and (4) an abrasion/scratch-resistant coating or a highly functional antireflective coating. Antireflective coatings are an important feature of photochromic lenses. They provide a functional quality for the lens-wearer and are esthetically pleasing (Guenther, 1985; Taniguchi and Seki, 1990). As with other photochromic lenses, all coatings which contain UV absorbers must be avoided on NISO and QISO lenses in order that the response of the photochromic is not significantly decreased.

In the early 1980s, American Optical (AO) commericalized, on a national level, a NISO photochromic lens. At that particular time, the eyewear industry expectations were that the AO plastic photochromic lens would be a glass photochromic lens look-alike in terms of activated hue and response performance. However, that was not the case and the lens was unsuccessful in the marketplace, and was later withdrawn.

Presently, two manufacturers are producing and commercially marketing plastic photochromics lenses in the US. All of these lenses utilize the indolinospiroxazine-type photochromics. More recently, Toray Industries (1989b) has disclosed the development of a plastic photochromic lens in which the photochromic is incorporated into the lens by the application of a hard coating layer. The specific class of photochromic used in these lenses has not been disclosed.

Rodenstock US Lens Division (1989) is currently distributing several different photochromic lenses, all apparently based upon the indo-linospironaphthoxazine photochromics. The first lens marketed was the Colormatic® in the progressive-plus design, and this exhibited a bleached brownish tint with a transmission of 75%. A special antireflective coating (Melzig, 1987) which enhances the life of the photochromic effect but does not attenuate the activating UV radiation is applied. This lens has been described by Rodenstock as exhibiting 'photocomfortability'. More recently, a similar photochromic lens, called Colormatic® Hard-Resin, has been marketed in standard single-vision, aspheric and progressive-plus lens designs. Similarly, this lens, without an antireflection coating, has also a brownish tint in the bleached state with a transmission of 75%. The UV activation and thermal bleach profiles at 10, 20 and 30°C are shown in Figure 2.18. The equilibrium response is 35% transmission at 20°C and 52% at 30°C. As with most photochromic lenses, both organic- and inorganic-based photochromics, the lens shows temperature dependency. The UV-activated lens spectrum shows a single-medium absorption band with a

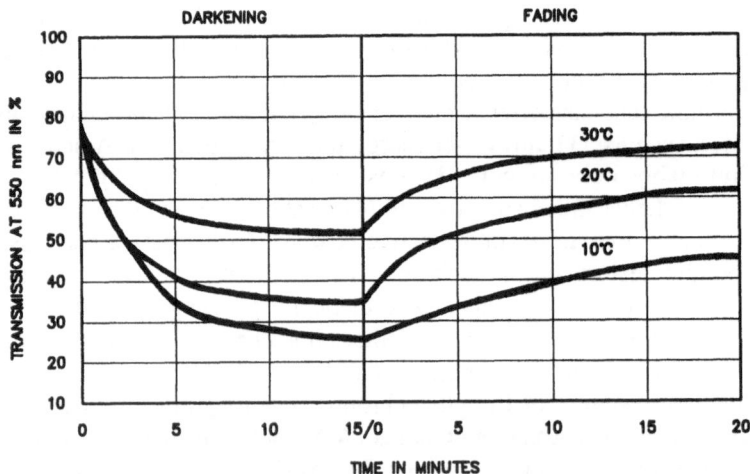

Figure 2.18 Photochromic performance of the Rodenstock Colormatic® lens.

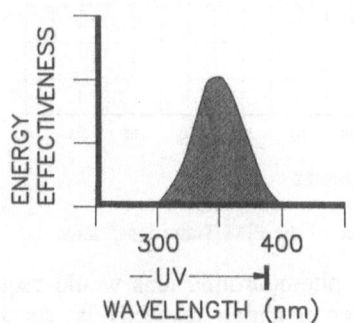

ENERGY EFFECTIVENESS

300   400

—UV———◄
WAVELENGTH (nm)

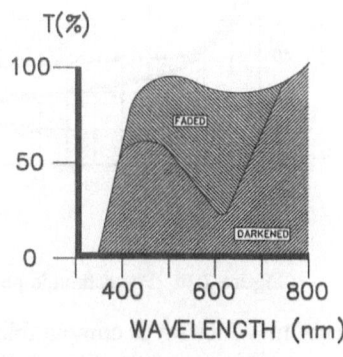

T(%)

100 —

50 —

0 —

FADED

DARKENED

400   600   800
WAVELENGTH (nm)

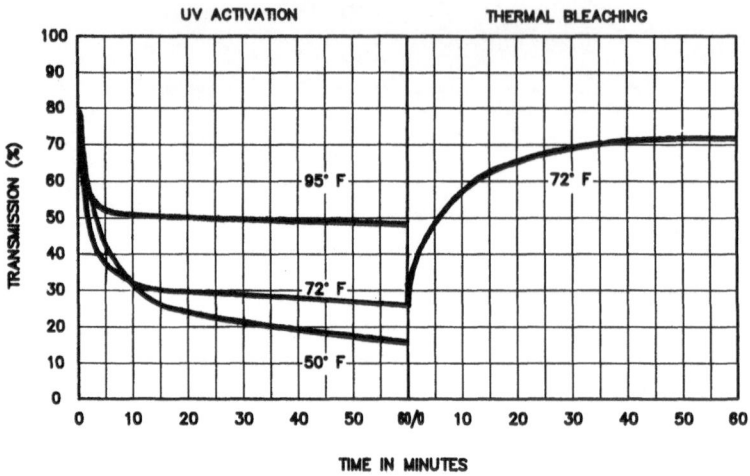

Figure 2.20 Photochromic performance of the PPG Transitions® lens.

correction, which in a comparable glass photochromic lens would require the center section of the lens to be lighter or darker respectively. As with most photochromic lenses, Transitions® offers nearly complete UV protection with greater than 95% UV absorption over the 295–400 nm range.

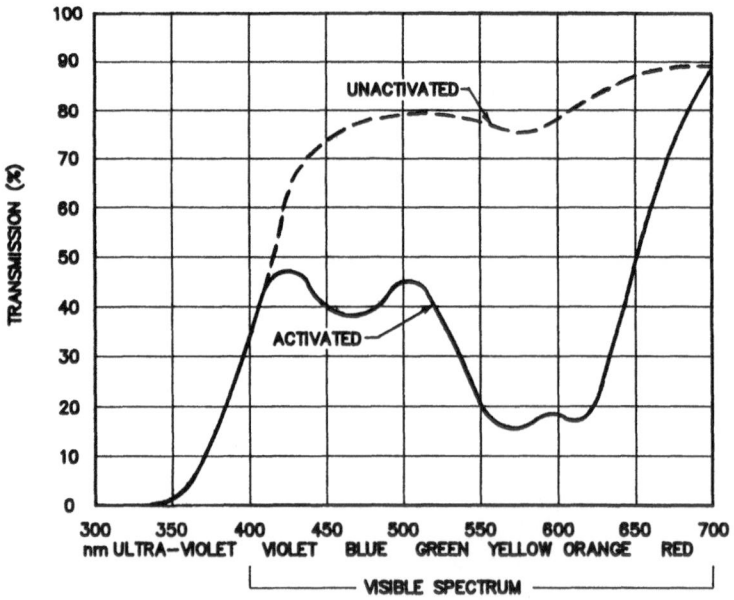

Figure 2.21 The unactivated and activated spectra for the PPG Transitions® lens.

References

Akashi, R. and Taniguchi, T. (1988a) Eur. Pat. Appl. 277,639 A2 to Toray Industries.
Akashi, R. and Taniguchi, T. (1988b) Jap. Pat. Appl. 63 308,014 A to Toray Industries.
Akashi, R., Taniguchi, T. and Jigyosho, S. (1988) Jap. Pat. Appl. 62 144,461 A to Toray Industries.
Aldoshin, S.M. and Atovmyan, L.O. (1987) *Mol. Cryst. Liq. Cryst.* **149**, 251.
Aoto, M., Nakamura, S., Maeda, S., Tomotake, Y., Matsuzaki, T. and Murayama, T. (1989) *MRS Int. Mtg Adv. Mats.* **12**, 219.
Armistead, W.H. and Stookey, S.D. (1964) *Science* **144**, 150; (1965) US Pat. 3,208,860 to Corning.
Arnold, G. and Vollmer, H.P. (1970) German Offen. 1,927,849.
Beattie, J.O. (1951) US Pat. 2,542,386.
Bercovici, T., Heiligman–Rim, R. and Fischer, E. (1969) *Mol. Photochem.* **1**, 23.
Bertelson, R.C. (1971) Photochromic process involving heterolytic cleavage. In *Photochromism — Techniques of Chemistry*, Vol III, ed. Brown, G.H. Wiley Interscience, London.
Bohne, C., Fan, M.G., Li, Z.J., Lusztyk, J. and Scaiano, J.C. (1990) *J. Chem. Soc., Chem. Comm.*, 571.
Brennan, P. and Fedor, C. (1989) *Plastics Compounding* Jan/Feb, 44.
Canon (1986) Jap. Pat. Appl. 61 175,081 A.
Casilli, N., Crisci, L., Renzi, F. and Rivetti, F. (1989) Eur. Pat. Appl. 316,980 Al to Enichem.
Chu, N.Y.C. (1983a) *Can. J. Chem.* **61**, 300.
Chu, N.Y.C. (1983b) German Pat. Appl. 3,310,388 A1 to American Optical.
Chu, N.Y.C. (1983c) UK Pat. Appl. 2,117,390 A to American Optical.
Chu, N.Y.C. (1984a) *Proceedings of the 10th IUPAC Symposium on Photochemistry, Interlaken, Switzerland.*
Chu, N.Y.C. (1984b) US Pat. 4,440,672 to American Optical.
Chu, N.Y.C. (1985a) *SPIE's 29th Annual International Technology Symposium Optical and Electroptical Engineering*, 562–03 (1985).
Chu, N.Y.C. (1985b) Eur. Pat. Appl. 134,633 A2 to American Optical.
Chu, N.Y.C. (1986a) Eur. Pat. Appl. 195,898 Al to American Optical.
Chu, N.Y.C. (1986b) *Solar Energy Materials* **14**, 215.
Chu, N.Y.C. (1987) US Pat. 4,699,473 to American Optical.
Chu, N.Y.C. (1988a) DOE/ST/1700-T1 to American Optical.
Chu, N.Y.C. (1988b) *Optical Materials Technology for Energy Efficiency and Solar Energy Conversion VII*, SPIE Vol. 1016 p. 152.
Chu, N.Y.C. (1988c) US Pat. 4,720,356 to American Optical.
Chu, N.Y.C. (1989) *Energy Res. Abstr.* **14(10)** No. 20059.
Clegg, W., Norman, N.C., Lasch, J.G. and Kwak, W.S. (1987) *Acta Cryst.* **C43**, 804.
Clegg, W., Norman, N.C., Flood, T., Sallens, L., Kwak, W.S. and Lasch, J.G. (1990) *Acta Cryst., Sect. C* (submitted for publication).
Corning Co. (1988) *Trade Literature*. Corning, New York.
Crano, J.C., Kwiatowski, P.L. and Hudritch, R.J. (1989) PCT Int. Appl. 8907278 A to PPG Industries.
Dateoka, Y. and Sagawa, T. (1987) Jap. Pat. Appl. 61 288,682 A to Nissan Motor.
Ferrara, M. (1990) *20/20* May, 58.
Flannery Jr., J.B. (1968) *J. Am. Chem. Soc.* **90**, 5660.
Guenther, K.H. (1985) *SPIE-Ophthalmic Optics* **601**, 76.
Heiligman-Rim, R. (1961) *J. Chem. Soc.*, 156.
Heiligman-Rim, R., Hirshberg, Y. and Fischer, E. (1962) *J. Phys. Chem.* **66**, 2465 and 2470.
Hirshberg, Y. (1950) *Compt. Rend.* **231**, 903.
Hirshberg, Y. and Fischer, E. (1953) *J. Chem. Soc.*, 629.
Hosoda, M. (1986) Eur. Pat. Appl. 186,364 A2 to Unitika.
Hovey, R.J. (1981) US Pat. 4,289,497 to American Optical.
Hovey, R.J., Chu, N.Y.C., Piusz, P.G. and Fuchsmann, C.H. (1980) US Pat. 4,215,010 to American Optical.

Hovey, R.J., Chu, N.Y.C., Piusz, P.G. and Fuchsmann, C.H. (1982) US Pat. 4,342,668 to American Optical.
Inoue, E., Kokado, H. and Kobayashi, H. (1968) *Kogyo Kagaku Zasshi* **71**, 1228.
Kawauchi, S., Saeda, S. and Yoshida, H. (1987) PCT Int. Appl. WO 8703874 to Showa Denko.
Kawauchi, S., Yoshida, H., Yamashina, N. and Ohira, M. (1990) *Bull. Chem. Soc. Jpn.* **63**, 267.
Kellmann, A., Tfibel, F., Dubest, R., Levoir, P., Aubard, J., Puttier, E. and Gugglielmetti, R. (1989) *J. Photochem. Photobiol., A: Chem.* **49**, 63.
Kido, T. (1988) Jap. Pat. Appl. 63 308,087 A to Pilot Ink.
Kluter, U., Hub, W. and Schneider, S. (1985) Time resolved vibrational spectroscopy. *Proc. Phys.* **4**, 152.
Kolc, J. and Becker, R.S. (1967) *J. Phys. Chem.* **71**, 4045.
Kureha Chem. Ind. (1988) Jap. Pat. Appl. 63 30,584 A.
Kwak, W.S. (1989) US Pat. 4,831,142 to PPG Industries.
Kwak, W.S. and Chen, C.W. (1989*a*) US Pat. 4,816,584 to PPG Industries.
Kwak, W.S. and Chen, C.W. (1989*b*) US Pat. 4,816,584 to PPG Industries.
Kwak, W.S. and Hurditch, R.J. (1987) US Pat. 4,637,698 to PPG Industries.
Kwak, W.S. and Hurditch, R.J. (1988) US Pat. 4,720,547 to PPG Industries.
Kwiatkowski, P.L. and Hunt, D.A. (1988) US Pat. 4,792,224 to PPG Industries.
Le Naour-Sene, L. (1981) US Pat. 4,286,957 to Essilor.
Machida, K., Saito, A. and Sakagami, J. (1990) Eur. Pat. Appl. 350,009 to Kureha.
Maltman, W.R. and Threlfall, I.M. (1987) Eur. Pat. Appl. 227,337 A2 to Pilkington Brothers.
Manfred, M. and Martinuzzi, G. (1984) Eur. Pat. Appl. 146,135 A to Rodenstock.
Matsui Chem. Ind. (1990) Jap. Pat. Appl. 2,034,686 A.
Matsuoka, S., Tanaka, T. and Kida, Y. (1988) Jap. Pat. Appl. 63 179,879 A.
Melzig, M. (1985) German Offen. DE 3,516,568 A1 to Rodenstock.
Melzig, M. (1987*b*) Ger. Offen. DE 3,525,891 to Rodenstock.
Melzig, M. (1989) Eur. Pat. Appl. 339,661 A2 to Rodenstock.
Melzig, M. (1990) Eur. Pat. Appl. 362,771 A1 to Rodenstock.
Mitsubishi Gas Chem. (1986) Jap. Pat. Appl. 61,005,910 A.
Muskat, I.E. and Strain F. (1945) US Pat. 2,379,250 to PPG Industries.
Nakajima, M., Iryo, T. and Mogami, T. (1988*b*) Jap. Pat. Appl. 63 267,786 A to Seiko Epson.
Nakajima, M., Iryo, T. and Mogami, T. (1988*a*) Jap. Pat. Appl. 63 267,785 A to Seiko Epson.
Nedoshivin, V. Yu., Lyubimov, A.V., Zaichenko, N.L., Marevtsev, V.S. and Cherkashin, M.I., (1989) Izv. Akad. Nauk SSSR, Ser. Khim, **11**, 2576.
Nissan Motor and Mitsubishi Kasei (1989) Jap. Pat. Appl. 1,180,536 A.
Nissan Motor and Mitsubishi Kasei (1990*a*) Jap. Pat. Appl. 2,034,541 A.
Ono, H. and Osada, C. (1970) Great Britain 1,186,987 to Fuji.
Ono, H. and Osada, C. (1971) US Pat. 3,562,172 to Fuji.
Ono, H., Osada, C. and Kosuge, K. (1971) US Pat. 3,578,602 to Fuji.
Optical Manufacturers Association (1990) *National Consumer Eyewear Study VI.*
PPG Industries (1989) *Trade Literature.* Pittsburgh, PA.
Reichenbacher, M. and Czerney, P. (1986) German Democratic Republic Pat. 238,611 A1.
Rickwood, M. and Hepworth, J.D. (1987) Eur. Pat. Appl. 245,020 A2 to Pilkington Brothers.
Rodenstock US Lens Division (1989) *Trade Literature.* Danbury, CT.
Rosenthal, F.S., Safran, M. and Taylor, H.R. (1985) *Photochem. Photobiol.* **42(2)**, 163.
Rossotti, H., (1983) *Colour.* Princeton University Press, Princeton, NJ, p. 118.
Sakagami, T., Machida, K., and Fujuii, Y. (1986) Jap. Pat. Appl. 61 236,521 A to Kureha.
Sakagami, T., Machida, K., Fujuii, Y., Arakawa, N., and Murayama, N. (1988) US Pat. 4,756,973 to Kureha.
Schneider, S.Z. (1987) *Phys. Chem. N.F.* **154**, 91.
Schneider, S. (1987*a*) *Z. Phys. Chem. N.F.* **154**, 91.
Schneider, S., Mindle, A., Elfinger, G. and Melzig, M. (1987*b*) *Ber. Bunsenges Phys. Chem.* **91**, 1222.
Sears, F.W. (1949) *Optics.* Addison-Wesley, Wokingham, p. 323.
Seiko Epson (1987) Jap. Pat. Appl. 62 226,134 A.
Sivadjian, J. (1968) *Bull. Soc. Chim. Fr.* 4829.
South Florida Test Service (1989) *Technical Yearly Summary Data.* Miami, FL.

Takuma, K., Kuroda, S. and Soga, H. (1988) Jap. Pat. Appl. 63 93,788 A to Mitsui ToatSu.
Tamaki, T. and Ichimura, K. (1988) *Proceedings of the Chemical Society of Japan 56th Spring Meeting, Koenyokoshu* II, 36.
Tanaka, T. and Kida, Y. (1987) Jap. Pat. Appl. 62 72,778 A to Tokuyoma Soda.
Tanaka, T., Matsuoka, S. and Kida, Y. (1987) Jap. Pat. Appl. 62 205,185 A to Tokuyoma Soda.
Taniguchi T. (1988) Jap. Pat. Appl. 63 141,973 A to Toray Industries.
Taniguchi, T. and Seki, T. (1990) US Pat. 4,904,525 to Toray Industries.
Taniguchi, T., Ohashi, K. and Ueno, T. (1989) Jap. Pat. Appl. 64 170,904 A to Toray Industries.
Tateoka, Y. Ito, M. Maeda, S., Kimura, M. and Mitsuhashi, K. (1987) PCT Int. Appl. WO 8700464 to Nissan Motor and Mitsubishi Kasei.
Tateoka, Y., Sagawa, T. and Kawasaki, M. (1988) Jap. Pat. Appl. JP 63 27,837 A to Nissan Motor and Unitika.
Tateoka, Y. Ito, M. Maeda, S. and Mitsuhashi, K. (1989a) Jap. Pat. Appl. 01 126,644 to Nissan Motor and Mitsubishi Kasei.
Tateoka, Y., Ito, M., Maeda, S., Mitsuhashi, K. and Murayama, T. (1989b) Eur. Pat. Appl. 313,941 A to Nissan Motor and Mitsubishi Kasei.
Tolar, H.R. (1974) US Pat. 3,877,798.
Toray Industries (1985) Eur. Pat. Appl. 171,909 A.
Toray Industries (1989a) Jap. Pat. Appl. 1,170,904 A.
Toray Industries (1989b) *Japanese High Technology Monitor* 7(17).
Toray Industries (1990) Jap. Pat. Appl. 2 042,084 A.
Tyer Jr., N.W. and Becker, R.S. (1970) *J. Am. Chem. Soc.* **92**, 1289.
Uhlmann, D.R., Snitzer, E., Hovey, R. J., Chu, N.Y.C. and Fournier, J.T. (1979) US Pat. 4,166,043 to American Optical.
Uhlmann, D.R., Snitzer, E., Hovey, R.J. and Chu, N.Y.C. (1983) US Pat. 4,367,170 to American Optical.
Unitika, K.K. (1986) Jap. Pat. Appl. 61 276,883 A.
Urbach. f., Rose, D.B. and Bonnem, M. (1972) *Institute Environment and Cancer*. Williams and Wilkins, Baltimore, MD, p. 476.
Welch, C. (1989) US Pat. 4,880,667 to PPG Industries.
Yamamoto, S. and Taniguchi, T. (1986a) Jap. Pat. Appl. JP 61 161,287 A to Toray Industries.
Yamamoto, S. and Taniguchi, T. (1986b) Jap. Pat. Appl. 61 233,079 A to Toray Industries.
Yamamoto, S. and Taniguchi, T. (1987) Jap. Pat. Appl. 62 33,184 A to Toray Industries.
Yamamoto, S. and Taniguchi, T. (1988) Jap. Pat. Appl. 63 275,587 A to Toray Industries.
Yamamoto, S. and Taniguchi, T.F. (1989a) Jap. Pat. Appl. 64 19,081 A to Toray Industries.
Yamamoto, S. and Taniguchi, T. (1989b) Jap. Pat. Appl. 64 52,783 A to Toray Industries.
Yamamoto, S. and Taniguchi, T. (1989c) PCT Int. Appl. WO 8907104 to Toray Industries.
Yamamoto, S. and Taniguchi, T. (1989d) Jap. Pat. Appl. 01 203,392 A to Toray Industries.
Yamamoto, S. and Taniguchi, T. (1989e) Jap. Pat. Appl. 01 207,292 A to Toray Industries.
Yoshitake, J., Munakata, Y., Tsutsui, T. and Saito, S. (1989) *Nippon Kagaku Kaishi* **1**, 127.

3 Fulgides and fulgimides — a promising class of photochrome for application

J. WHITTALL

3.1 Introduction to fulgides and fulgimides

Stobbe (1893) reported that the attempted Claisen condensation of diethyl succinate with acetone did not yield the expected ethyl heptan-4,6-dionate **1** but gave instead the unsaturated half-ester **2** (Figure 3.1).

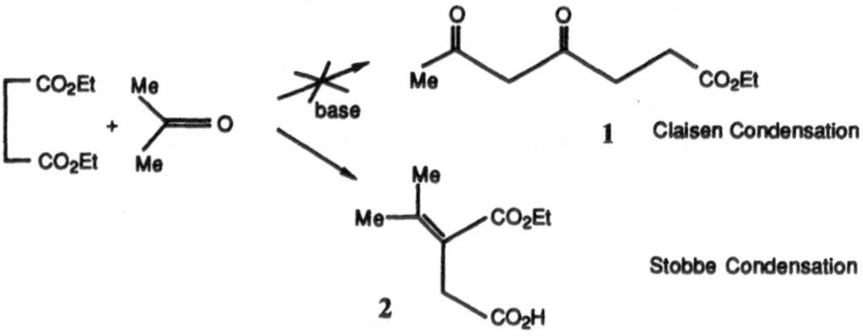

Figure 3.1 The Stobbe condensation.

Later work by Stobbe and co-workers demonstrated that the reaction was a general one, and a wide range of aldehydes and ketones were found to condense with diethyl succinate in the presence of base to give unsaturated half-esters. It was also established that esterification of these derivatives gave the substituted diethyl methylenesuccinates **3** which underwent a second Stobbe condensation to give substituted bis-methylene succinic half-esters **4**. The latter could be hydrolysed to the diacids **5** which on dehydrative cyclization gave the bis-methylene succinic anhydrides **6**. Stobbe named these compounds 'fulgides' (from the Latin *fulgere*, to glisten and shine) as they often crystallized as bright, shiny crystals. This work was published over the period 1893–1911 and culminated in a review after a series of papers by Stobbe and co-workers (1911) from which details of this work and further references may be obtained.

Subsequently, the related imide derivatives **7** were named 'fulgimides' because of their relationship with succinimides (Heller *et al.*, 1968).

The Stobbe condensation was thoroughly investigated by Johnson *et al.* (1950), who elucidated the mechanism. They demonstrated the formation of an intermediate lactonic ester **8** which explained the regiospecific formation of the half-ester by irreversible intramolecular base-induced elimination of the lactonic ester to give the half-ester salt **9** (Figure 3.2).

The Stobbe condensation was reviewed by Daub and Johnson (1951) and its scope and limitations were discussed. It can be seen that aldehydes and unsymmetrical ketones can give rise to two isomeric half-esters, and for unsymmetrical ketones this is what occurs. Most aromatic aldehydes have been shown to form the (E) half-ester exclusively. Heller and Szewczyk (1974) suggested a mechanism which involved 'orbital overlap control' in the irreversible formation of the half-ester.

Whilst strict IUPAC nomenclature for these compounds defines them as derivatives of furan-2,5-dione, they are normally named as substituted

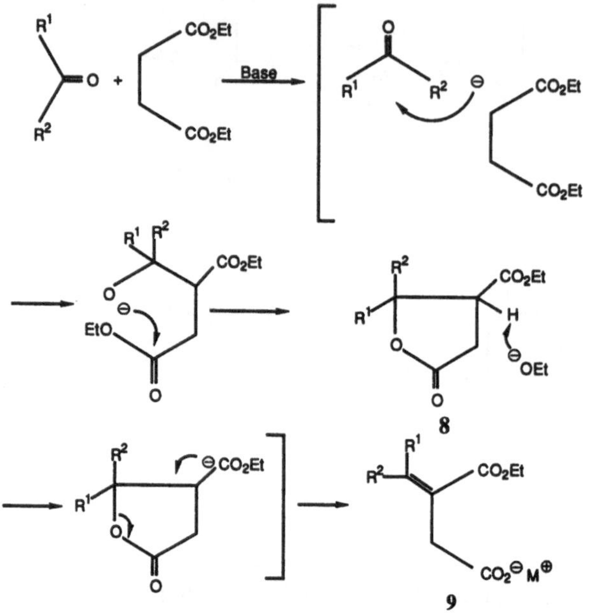

Figure 3.2 Mechanism of the Stobbe condensation. (From Johnson *et al.*, 1950.)

bis-methylenesuccinic anhydrides or fulgides. Greek letters are used to describe the location of substituents as shown below.

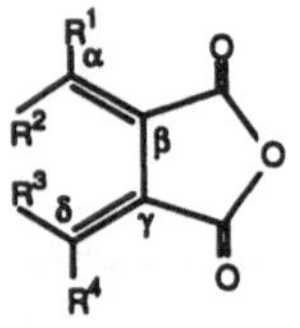

The assignment of stereochemistry as *cis* or *trans* can be ambiguous, and so stereochemistry is indicated as (E) or (Z) based on the sequence rules using the Cahn–Prelog–Ingold convention.

3.2 Aryl fulgides

3.2.1 Introduction

The investigations that led to the synthesis, characterization and elucidation of photochemistry of fulgides were carried out on phenyl and substituted phenyl fulgides. Although these seem to have less potential in practical applications than some heteroaromatic substituted fulgides discussed later in this chapter, a review of the properties of aryl fulgides gives an essential insight into the properties of fulgides.

3.2.2 Spectroscopic properties

Fulgides and fulgimides substituted with four different groups R^1–R^4 in 6 or 7 can exist as four geometrical isomers [(E,E); (E,Z); (Z,E); (Z,Z)]. Some early workers in this field (Chakraborty et al., 1966; Brunow and Tylli, 1968; Swoboda et al., 1967; Abdel-Wahhab and El-Assal, 1968; Harper et al., 1970; Abdel-Wahhab and Rayes, 1971) were unable to accept the (E,E) configuration for bis-aryl fulgides 6 (R^4, R^2 = H, R^1, R^3 = Ar), claiming that steric overcrowding would prevent this isomer from being formed, even though Stobbe (1911) had already synthesized the tetraphenyl fulgide 6 (R^{1-4} = Ph) and had demonstrated that it was a stable crystalline compound and therefore it was known that it was possible to have aryl groups overlapping and forming overcrowded fulgide molecules.

The following spectroscopic techniques have been used to determine the structure, stereochemistry and properties of fulgides.

3.2.2.1 X-ray crystallography.
Cohen et al. (1970) prepared the bis-(p-methoxyphenyl) fulgide 10 and used X-ray crystallography to demonstrate that the fulgide was in the (E,E) configuration with the aryl groups twisted by approximately 30° to the plane of the anhydride ring. This work was confirmed by Boeyens et al. (1988a), who also showed that the yellow prisms (m.p. 168–168.5°C) had the two methoxy groups oriented in the same sense but that on refluxing a solution of this isomer in acetone a rotamer was obtained as orange prisms (m.p. 176–176.5°C) with the two methoxy groups oriented in an opposed sense (11). Both rotamers in solution had identical UV absorption and nuclear magnetic resonance (NMR) spectra because of rapid interconversion in the solvated state, but in the crystalline state a high-energy barrier restricted rotation of the methoxy groups.

10   11

Boeyens *et al.* (1988b) also carried out an X-ray crystallographic study on the bis-(3,4-dimethoxyphenyl) fulgide **12** (Ar = 3,4-dimethoxyphenyl) and found that the aryl rings were inclined at an angle of 50°. No unusual bond lengths were observed but the bond angles at the olefinic trigonal carbon atoms were found to be strained, as shown in **12**.

12

In order to keep the internal angle at the carbonyl groups close to the 108° required for a planar pentagon, the external angle of the methylene groups opens up to 137.5° and the carbonyl groups are opened up making an angle of 130.6° to the ring for the same reason. The aryl groups also have the angles to the methylene group double bonds opened up to 134.2° but there is no obvious explanation for this.

Thus, X-ray crystallography has been used to establish the geometry of the fulgide isomers and to demonstrate the strain in the molecular structure which results in restricted rotation about single bonds and extended bond angles.

3.2.2.2 Proton nuclear magnetic resonance spectroscopy. The difficulty in accepting the (E,E) geometry of bis(aryl) fulgides resulted in incorrect assignments of proton NMR signals by some early workers in this area (section 3.2.2). However, Heller and Hart (1972) established that proton NMR spectroscopy was the most convenient method for determining the stereochemistry of fulgides. The two major factors which allowed this technique to be used were the strong deshielding effect of the carbonyl groups on any groups *cis* to them and a strong shielding effect on groups underlying an aryl group *trans* to a carbonyl group. Some examples to demonstrate these effects are given in Figure 3.3.

Ilge and Schutz (1984) published data on over 30 fulgides which confirmed these magnetic anisotropic effects, and they also demonstrated that steric effects alter the magnitude of these effects by twisting bond angles. This result could be confirmed by changing solvents (from $CDCl_3$ to $C_6 D_6$) and investigating the solvent shifts so induced.

3.2.2.3 Carbon-13 NMR spectroscopy. Ilge and Paetzold (1984a) reported the carbon-13 chemical shifts for a range of phenyl and substituted phenyl fulgides. Comparison of these results with those of related compounds gives some insight into the structure of the fulgides.

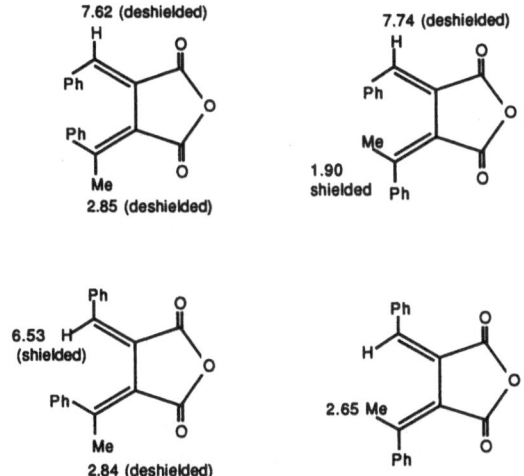

Figure 3.3 Proton NMR data for selected fulgides [δ values (ppm) for CDCl$_3$ solutions with tetramethylsilane internal standard].

For (E)-cinnamic acid **13** there is a strong polarization of the double bond which transmits the conjugation between the electron donor phenyl group and the electron acceptor carboxylic acid group which results in the C$_\beta$ atom being strongly deshielded relative to the C$_\alpha$ atom (Δδ = 30.1 ppm). The necessity for the phenyl group to be present for such polarization to be seen is demonstrated by ethyl acrylate **14** where the C$_\alpha$ and C$_\beta$ atoms have almost identical chemical shifts (Δδ = 3.9 ppm) (Modro *et al.*, 1990).

The related diphenyl fulgide **15** had similar chemical shifts to cinnamic acid **13** but the deshielding was slightly less (Δδ = 20.2 ppm), which was interpreted as a measure of the reduced π polarization in the fulgide in comparison with cinnamic acid. There was also a good linear correlation between the carbon-13 chemical shifts and the net atomic charge density as calculated by the MNDO method for the fulgide of **15**.

The carbon-13 chemical shift differences were also used as an estimate of the relative rotation of the aryl rings in a number of phenyl fulgides with different substituents elsewhere in the molecule by using these measurements as an estimate of the π-bond polarization. The effects on π-bond polarization of *para*-nitro or *para*-methoxy substituents on the phenyl groups were also estimated (Ilge *et al.*, 1984a).

3.2.2.4 Mass spectroscopy.   The fragmentation patterns and high-energy rearrangement mechanisms have been reported by Ilge and Paetzold (1984c).

3.2.2.5 Infrared spectroscopy.   Ilge *et al.* (1984) described the infrared absorption spectra for a range of fulgides. There were characteristic absorptions for the succinic anhydride moiety at 1816–1831 cm^{-1} (asymmetric stretch) and 1765–1776 cm^{-1} (symmetric) and for the butadiene moiety at 1604–1644 cm^{-1}.

3.2.2.6 Ultraviolet spectroscopy.   Many different aryl fulgides have been studied, and the changes in the electronic absorption spectra resulting from changes in substituents and structural isomerizations have been elucidated.

(a) *Effect of number of aryl groups.* There is a bathochromic shift in the longest wavelength absorption band for every phenyl group attached to the α or δ positions of fulgides. This is clearly shown by the fulgides **16 a–e** where there is a bathochromic shift of between 28 and 55 nm as the methyl groups are replaced by phenyl groups, as shown in Figure 3.4.

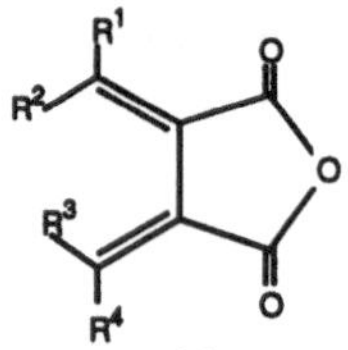

		λ_{max} nm
16a $R^1 - R^4$ = Me		276
16b R^1, R^2, R^4 = Me, R^3 = Ph		312
16c R^1R^4 = Me, R^2R^3 = Ph		340
16d R^1 = Me, R^2, R^3, R^4 = Ph		375
16e $R^1 - R^4$ = Ph		420

Figure 3.4 Bathochromic shifts due to increasing number of phenyl groups in fulgides.

(b) *Effect of α-alkyl substituents.* Replacement of an α-hydrogen *cis* to the carbonyl group by an α-methyl substituent results in a hypsochromic shift and has a hypochromic effect on the longest wavelength absorption band of fulgides. This is because steric interactions between the methyl and carbonyl oxygen, cause more twisting of the chromophore (see Figure 3.5), which results in less polarization of the chromophore.

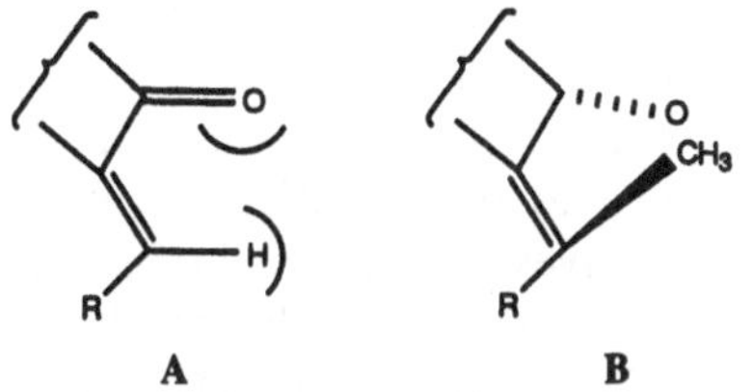

Figure 3.5 Increased steric interactions in α-methyl fulgides (B) relative to analogues (A).

(c) *Other substituents.* The introduction of other substituents can have very large effects on the electronic absorption spectra of fulgides. A good example is the fluorenylidene group in fulgides, which causes very large bathochromic shifts relative to an isopropylidene group in fulgides (100–150 nm). The bis-fluorenylidene derivative **17** forms deep-black crystals which give violet solutions in organic solvents (λ_{max} = 502 nm in toluene) (Goldschmidt *et al.*, 1957).

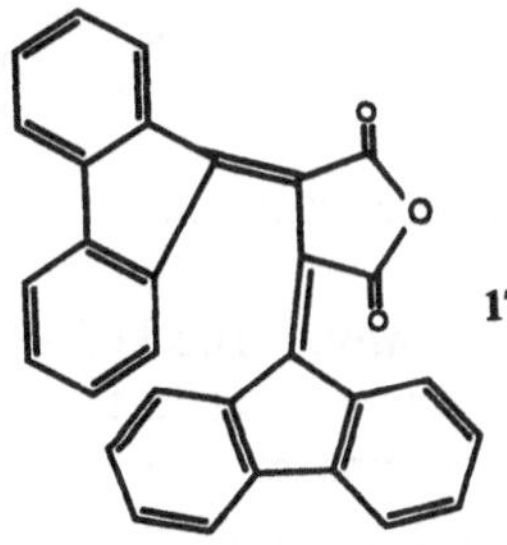

3.2.3 Photochromism

Fulgides are normally yellow/orange crystalline compounds which are photochromic in crystals, solutions, polymer matrices and glasses, changing to a deeper colour on exposure to UV light. The colour can be reversed either thermally or with white light. Stobbe first reported that fulgides were photochromic but failed to explain the phenomenon adequately, suggesting that it was either a crystal effect or that E–Z isomerizations were responsible for the observations. The fulgide **10** has been shown to undergo minor colour changes (Cohen *et al.*, 1970) on exposure to UV radiation because of photo-equilibrium of the pale (E,E) isomer to a photostationary-state equilibrium mixture with the deeper coloured (E,Z) and (Z,Z) isomers, but this is not a useful photochromic effect and these photochemical isomerizations are often unwanted side reactions that reduce efficiency in the desired photochromic reaction. Similar results have been reported by the same authors (Cohen *et al.*, 1977) for other related fulgides.

Santiago and Becker (1968) suggested that the photochromism was a molecular phenomenon due to photocyclization to the 1,8a-dihydronaph-thalene (1,8a-DHN) derivatives, which could be oxidized by molecular oxygen to form naphthalene derivatives (Figure 3.6). Heller *et al.* (1968) and Heller and Hart (1972) demonstrated that the mechanism was a con-certed conrotatory electrocyclic ring closure in accordance with the Woodward–Hoffman selection rules.

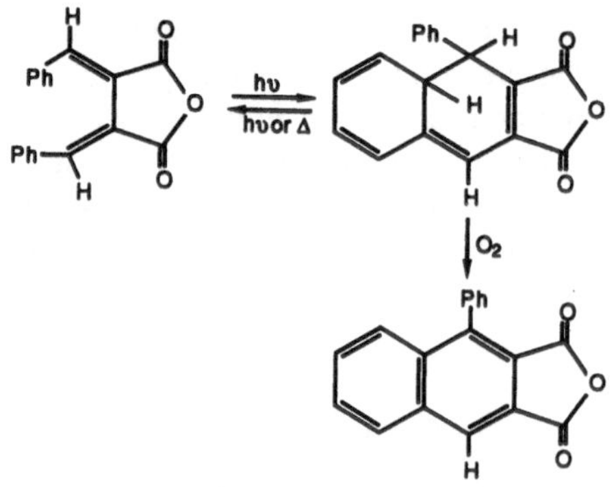

Figure 3.6 Reactions of aryl fulgides when irradiated in air.

3.2.3.1 Chromophore structure. The UV spectrum of dimethyl(E,E)-bis-benzylidene succinate **18** was similar to that of methyl *trans*-cinnamate **19** and did not resemble the spectrum of 1,4-diphenylbutadiene (Freudenberg and Kempermann, 1957) which indicates that compound **18** has two largely independent chromophores.

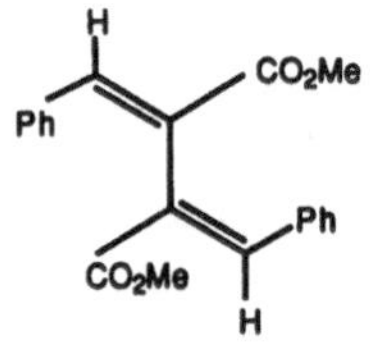

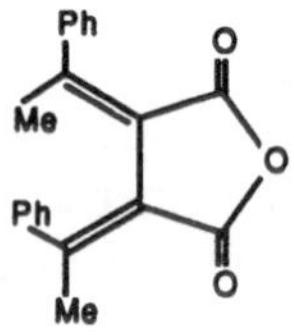

Heller and Szewczyk (1974) demonstrated that the UV spectrum of (E,Z)-bis-phenyl fulgide **20** was almost identical with a UV spectrum of a 1:1 mixture of the (E,E) and (Z,Z) isomers of the same fulgide. This evidence indicates that the fulgide system has two largely independent cinnamic anhydride-type chromophores, as shown in Figure 3.7 (Heller *et al.*, 1986a).

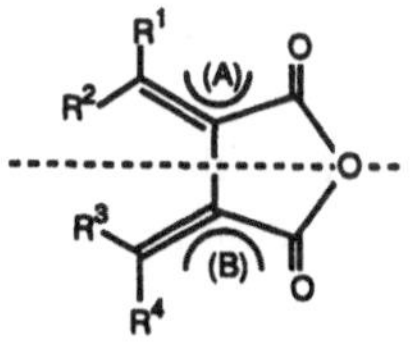

Figure 3.7 The two independent cinnamate chromophores of fulgides.

It has been suggested (Heller *et al.*, 1986a) that absorption of light by one chromophore then allows photocyclization onto an aryl group attached to the other chromophore to occur but that it does not allow photocyclization onto an aryl group attached to the chromophore which absorbs the photon.

This proposal was used to explain the observation that the fulgides **21a** and **22a, b** were markedly photochromic (deep-blue colours being observed on irradiation) while the fulgide **21b** was so poorly photochromic that it was difficult to observe any colour change on irradiation under the same conditions.

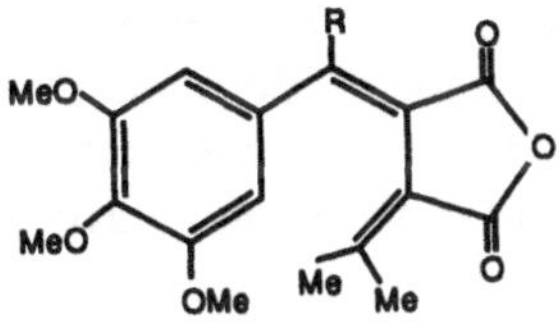

21a R=Me
21b R=H

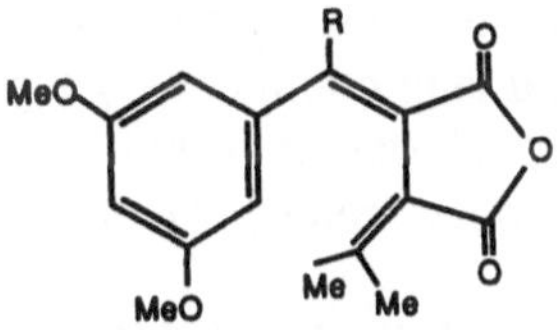

22a R=Me
22b R=H

For **21b**, the chromophore for which the excited state can be represented as the dipolar structure **23** dominates, and this causes E–Z isomerizations about the double bond of the absorbing chromophore to be the main photoreaction, while in other cases excited states represented by **24** are more favoured, and this allows photocyclization onto the aryl group to give the coloured 1,8a-DHNs.

23   24

Further evidence for this comes from the lactones **25** which on irradiation yielded photoproducts from the intermediate **26** which came from exclusive photocyclization onto the aryl ring opposite the carbonyl group. It seems likely that a similar mechanism to the photocyclization of fulgides operates in this case (Strydon, 1974).

25   26

3.2.3.2 Mechanistic aspects of the photochemistry of aryl fulgides. The fundamental photochemistry of aryl fulgides has been elucidated by two series of studies, one by Becker and co-workers, and a second initiated by Paetzold and later continued by Ilge and co-workers. Both groups agree that the singlet π,π^* state is responsible for the photocyclization reaction of fulgides, and other routes which cause energy dissipation have also been elucidated.

Santiago and Becker (1968) first suggested that the singlet state was responsible for the photocolouration as no triplet sensitization was possible and no phosphorescence was detected. Later studies have confirmed this hypothesis, as no triplet quenching of the photoreaction has been observed (molecular oxygen, ferrocene) and no triplet transients have been detected by either time-resolved absorption spectroscopy or electron spin resonance (Ilge *et al.*, 1984b, 1986). Internal or external heavy-atom effects could not induce intersystem crossing (Ilge and Paetzold, 1984).

Ilge and Paetzold (1984b) found that only simultaneous E–Z isomerization about both double bonds occurs when fulgide triplet states are formed in sensitization experiments. Common triplet states for (E,E) and (Z,Z) isomers and for both (E,Z) isomers were proposed, and the conversion of (E,E) or (Z,Z) isomers into either of the (E,Z) isomers did not occur in the triplet manifold (Figure 3.8).

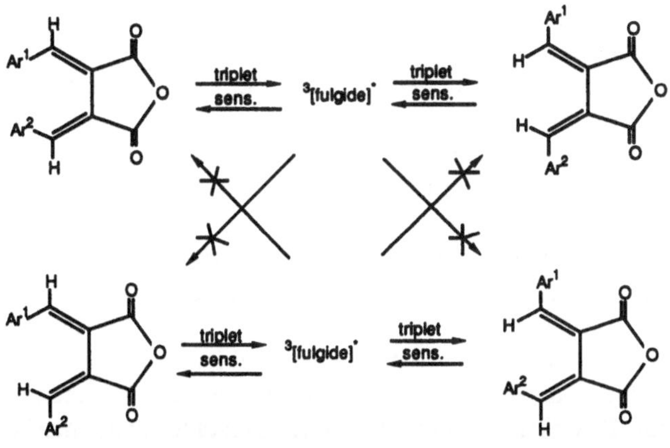

Figure 3.8 Senstized triplet photoreactions of fulgides.

Lenoble and Becker (1986) reported that photocyclization and E–Z isomerization occurred in less than a nanosecond for a range of fulgides. Ilge *et al.* (1986, 1987), using picosecond resolution, found that these processes were ultrafast and the photoproducts were formed in the pulse duration. The four competing deactivation processes from the singlet manifold were E–Z isomerization, photocyclization, internal conversion and fluorescence.

The E–Z photoisomerization reactions for the fulgide **27** were studied by Ilge (1986) because the electronic effects of the *para*-methoxy substituents on the phenyl groups made the quantum yield of photocyclization very low (< 10⁻⁵).

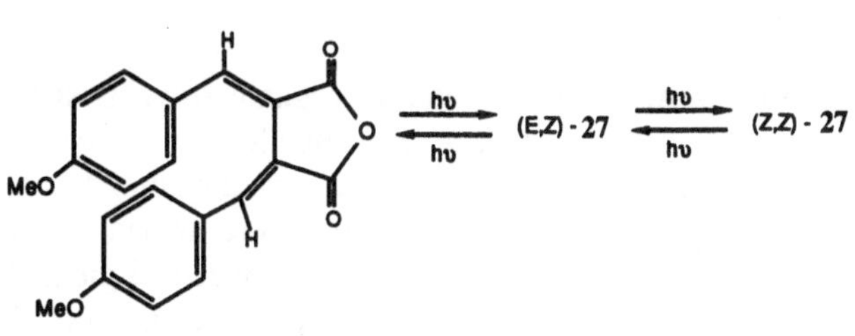

The quantum yield for isomerization of the (E,E)-fulgide **27** into the (E,Z)-fulgide **27** was independent of temperature (110–293 K), wavelength of irradiation (313–436 nm) and solvent (toluene, acetonitrile, cyclohexane) and had a value of 0.28. The quantum yields for the isomerization of the (E,Z)-fulgide **27** into either the (E,E)-fulgide **27** or the (Z,Z)-fulgide **27** were dependent on wavelength but were independent of temperature. The variation with wavelength of this quantum yield could be ascribed to different reaction routes from higher excited singlet states (S_2, etc.). For the photoisomerization to occur, rotation to a perpendicular state after absorption of a photon must occur, and the lack of temperature dependence and ultrafast nature of this process indicates there is no thermal activation barrier to this rotation for these more overcrowded isomers. However, the less overcrowded (Z,Z)-fulgide **27** has a quantum yield for photoisomerization to the (E,Z)-fulgide **27** that decreases as the temperature is decreased, and below 130 K weak fluorescence was also observed ($\Phi_f = 2 \times 10^{-4}$ at 77 K); hence there is a thermal activation barrier to the photochemical isomerization in this isomer.

The temperature dependence of the absorbance spectra of the three isomers of the fulgide **27** was also reported. The (Z,Z) isomer had a marked increase in absorbance at low temperature and this was because of a reduction of the torsion angle of the aryl ring by thermal processes at low temperatures. This effect was much less marked for the (E,E) and (E,Z) isomers of the fulgide **27** as the molecular overcrowding prevents reduction of the torsion angle because of steric interactions.

The photochemistry of a series of fulgides **28a–c, 29** and **30a–c** was determined by Paetzold and Ilge (1984) and the quantum efficiencies for the various processes determined.

28a R=H, **b** R=Me, **c** R=Ph   **29**   **30a** R=H, **b** R=Me, **c** R=Ph

At room temperature, these fulgides in organic solvents showed no sign of any luminescence and there was no evidence for any intersystem crossing. The quantum yield for E → Z isomerization varied from 0.40 for the fulgide **29** to 0.26 and 0.18 for the slightly more overcrowded fulgides (**28a, b** respectively), and was $< 10^{-3}$ for the highly crowded fulgides **30a, b**. This type of isomerization is not possible for fulgides **28c** and **30c**. The quantum yields for photocyclization were < 0.1 for all these fulgides but was highest for fulgides **28b, c, 29** and **30b** (0.084, 0.060, 0.046 and 0.01 respectively) and was $< 10^{-3}$ for the other fulgides (**28a, 30a, c**).

These results show that internal conversion is the major deactivation route (i.e. it accounts for 99% or more of the photons absorbed in fulgides **30a–c** and for 55% of the photons absorbed for the fulgide **29**) and that this deactivation process increases with overcrowding.

Torsion around the C_α-aryl ring single bond or collapse of the perpendicular excited state back to the starting isomer of the fulgide were suggested as possible routes for these processes.

The S_1 lifetime of the fulgide **28b** was shown to be 6 ps (by decay of the $S_1 \rightarrow S_n$ absorption) and the rise time of the 1,8a-DHN formed by photocyclization reaction was also nearly 6 ps.

There was no temperature dependence on the cyclization quantum yield, and so this reaction also does not have a thermal activation barrier.

It has been suggested that the mechanism has some similarities with that of the first step of aromatic electrophilic substitution because of the activating effect of electron-donating groups *ortho* or *para* to the position where cyclization occurs on the aromatic ring. This is exemplified by the fulgide **31b**, which has a quantum yield for photocyclization of 0.076, while the fulgide **31a** has a quantum yield of < 0.01 under the same conditions (toluene, room temperature, irradiation with 366 nm light) (Heller *et al.*, 1986a).

31a R=H, b R=OMe   32a R=H, b R=OMe

It can also be seen that the methoxy groups of 1,8a-DHN **32b** are in conjugation with the chromophore and cause this compound to be blue (λ_{max} 568 nm) compared with the orange–red for **32a** (λ_{max} 455 nm).

3.2.4 Fatigue resistance

All these fulgides gave 1,8a-DHNs that were not suitable for most applications because they reversed thermally to the fulgide by a disrotatory ring-opening process and also underwent a range of irreversible reactions which included thermal [1,5]-hydrogen shifts, photochemical [1,3]-hydrogen shifts and oxidation to naphthalenes. These reactions were all due to the

8a-hydrogen in the 1,8a-DHN, and replacement of this by a methyl group (as in the 1,8a-DHNs **34a, b**) gave red–purple 1,8a-DHNs which were thermally stable and did not undergo thermal ring opening, methyl shift or ethane elimination reactions (Heller and Megit, 1974). The lack of thermal disrotatory ring opening was attributed to steric interaction between the pseudoequatorial 1-methyl and the pseudoaxial 8a-methyl group which enter a direct collision course in the allowed disrotatory ring-opening mode. This interaction is absent in the conrotatory mode and the photochemical ring openings of the 1,8a-DHNs **34a, b** to fulgides **33a, b** with visible light occurs with good quantum efficiencies. Unfortunately, the conversion into the 1,8a-DHNs at the photostationary state on irradiation with 366-nm light was low.

33a R=H, b R=OMe   34a R=H, b R=OMe

3.3   Fulgides with heteroaromatic groups — fatigue-resistant fulgides

3.3.1   (E)-α-2,5-dimethyl-3-furylethylidene(isopropylidene)succinic anhydride

The aryl fulgides described in the previous section either had good photochromic properties with poor fatigue resistance (mainly because of hydrogen shifts or oxidative aromatization) or had poor photochromic properties. The major breakthrough in the development of fulgides which possess both good photochromic properties and fatigue resistance came with the synthesis of the 3-furyl fulgide **35** by Heller *et al.* (1981).

35   36

The potentially practical usefulness of this fulgide has resulted in extensive studies of its properties, photoreactions, photophysics and chemical transformations and the synthesis of many closely related compounds to evaluate the potential for new technologies.

3.3.1.1   Ultraviolet/visible spectroscopy and photochemistry. The fulgide **35** was a pale-yellow compound (λ_{max} 343 nm in toluene solution) which is conveniently close to the 366-nm resonance line of the mercury lamp. Irradiation with this light source caused quantitative photocyclization exclusively onto the 2 position of the furyl ring (rendering it unnecessary to protect the 4 position of the furyl ring by a methyl group) to give the

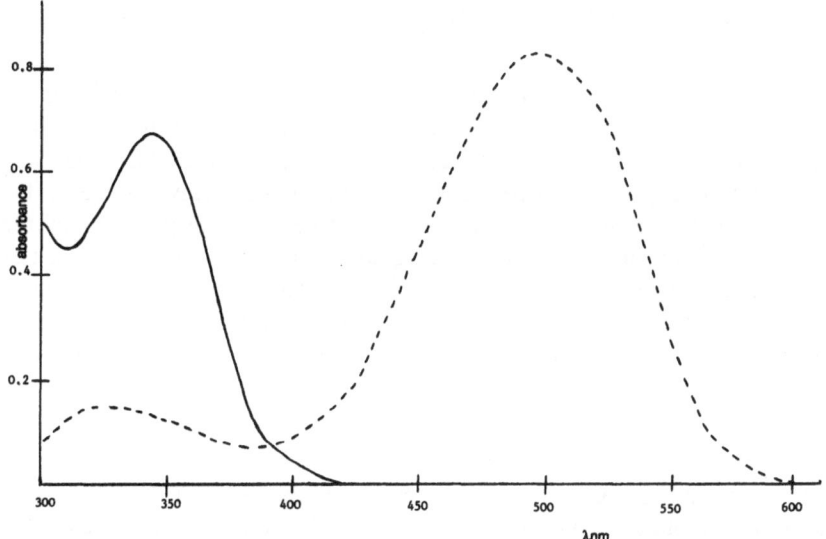

Figure 3.9 Absorption of the fulgide with structure 35 (solid line) and the DHBF with structure 36 (dashed line) (10^{-4} M solutions in toluene).

deep-red 7,7a-dihydrobenzofuran (7,7a-DHBF) **36** which has an absorption minimum in the near-UV region. This can be clearly seen in the UV/visible spectra of **35** and **36** shown in Figure 3.9.

Heller and Langan (1981) reported that the quantum yield for colouring of the fulgide **35** to the 7,7a-DHBF **36** in toluene solutions was 0.20. This value was independent of wavelength (313–366 nm range) and temperature (10–40°C temperature range) but did vary slightly (± 10%) for solutions in a variety of common organic solvents. Ulrich and Port (1989) studied the photoreaction in 1-μm thin films (formed by vapour deposition onto quartz surfaces) and found the same quantum yield for colouring in the solid state as for solution in organic solvents. They demonstrated that this was independent of temperature down to 50 K. This indicates that photocyclization is an ultrafast process with no barrier to cyclization (similar to the results discussed in section 3.2.3 for phenyl fulgides).

The visible light-induced ring opening of the red 7,7a-DHBF **36** back to the original fulgide has been shown to exhibit a remarkable linear dependence on both the wavelength of irradiation (within the longest wavelength absorption band) and temperature. The variation of this quantum yield with wavelength for a toluene solution at 21°C is given in eqn (1).

$$\Phi = 0.179 - 2.4 \times 10^{-4} \ (\lambda \ nm) \tag{1}$$

One explanation for these observations is that there is an activation energy barrier to the reaction (that can be overcome more easily from higher thermal energy levels in the excited state) from which the ring opening is occurring.

3.3.1.2 Proton NMR spectroscopy. The proton NMR absorptions for the fulgide **35** have been assigned by comparison with other fulgides (Heller *et al.*, 1981) and by selective deuteration (Heller and Oliver, 1981), but only arbitrary assignments could be made for some of the methyl signals in the 7,7a-DHBF **36**. Kaptein *et al.* (1986) confirmed the assignments of the proton NMR absorptions for the fulgide **35** using nuclear Overhauser enhancement (NOE) and then used a spin coherence transfer experiment in the photochemical reaction to give a two-dimensional proton NMR spectrum where the cross-peak connectivities give the assignments for the 7,7a-DHBF **36**.

3.3.1.3 X-ray crystallography and theoretical studies. Yoshioka *et al.* (1989) reported the X-ray crystal structures of the (E) and (Z) isomers of the fulgide **35** but could not obtain the 7,7a-DHBF **36** as crystals. The double bonds of the *cis*-butadiene part of the structure are slightly longer than for *cis*-1,3-butadiene, while the single bond is slightly shorter. The molecule is twisted (torsional angles −39.6°, −16.5°) and the succinic anhydride portion deviates from a planar system. The distance between the furan 2 carbon and the α^1 carbon (which are the two carbons which bond to form the six-membered ring during photocyclization) is 3.44 Å.

Yoshioka and Irie (1989) used MNDO-SCF semiempirical molecular orbital calculation and an *ab initio* (STO-3G) method and concluded that the LUMO mainly consists of the π^* antibonding LUMO of the *cis*-butadiene and that the π,π^* excited singlet state was responsible for the photoisomerizations. They also concluded that the DHBF **36** is of lower energy than the (E)-fulgide **35**.

A simple theoretical description of the colouration reaction and E–Z isomerizations in the π,π^* singlet state has been presented by Ilge and

Colditz (1990). The results are interpreted in terms of II-bond orders and free-valence indices for the bond-forming atoms calculated by the Pariser–Parr–Pope method.

Huang *et al.* (1984) studied the photoreaction of the 7,7a-DHBF **36** back to the fulgide **35** using 514.5-nm light and detected a transition state which decayed back to the 7,7a-DHBF **36** and the fulgide **35** with specific rates of 22.43/ms and 1.11/ms respectively. Hence two different reaction pathways in the photocolouration of **35** and the photodecolouration of **36** are involved.

3.3.2 Structural modifications of fulgides with heteroaromatic groups

Many structural modifications have been carried out to see if the photochromic response can be altered to make it more suitable for various applications. These have included alteration of the substituents on the furan ring, alteration of the substituents on the α positions, the synthesis of fulgimides, and the use of heterocyclic rings other than furan.

3.3.2.1 Variation of the α-alkyl substituent. Heller *et al.* (1981) reported that the fulgide **37b** was strongly photochromic when irradiated with 366-nm light, while the fulgide **37a** (where a hydrogen replaces the methyl in the α position) only underwent E–Z photoisomerization when irradiated under similar conditions but did not give any photocyclization to the coloured 7,7a-DHBF **38a**.

37 a R=H b R=Me c R=Et
    d R=^{n}Pr e R=^{i}Pr f R=^{n}C$_{17}$H$_{35}$

38 a R=H b R=Me c R=Et
    d R=^{n}Pr e R=^{i}Pr f R=^{n}C$_{17}$H$_{35}$

Hibino and Ando (1986a) synthesized the fulgide **37f** which can be used to construct multilayer Langmuir–Blodgett films which have photochromic properties similar to the photochromic properties observed in organic solvents.

Further studies by Kurita *et al.* (1988, 1990a) into the effect of the larger alkyl substituents in fulgides **37c–e** on the quantum yield for colouring demonstrated that this increased with increasing size of the α substituent R. These substituents had only small effects on the quantum yields for bleaching and on the absorption spectra of both the fulgides and the 7,7a-DHBFs **37a–e** as shown in Tables 3.1 and 3.2.

Table 3.1 Spectroscopic data for the fulgides **37a–e** (chloroform solutions).

	Fulgide				
	37a	**37b**	**37c**	**37d**	**37e**
λ_{max} (nm)	—	347	349	348	347
ε_{max} (dm³/mol cm)	—	6780	6690	6210	4080
Φ_{col}	0.00	0.2	0.34	0.45	0.62

Table 3.2 Spectroscopic data for the 7,7a-DHBFs **38a–e** (chloroform solutions).

	7, 7a-DHBF				
	38a	**38b**	**38c**	**38d**	**38e**
λ_{max} (nm)	Not	510	510	510	510
ε_{max} (dm³/mol cm)	formed	9690	10270	9590	9320
Φ_{rev}		0.035	0.027	0.035	0.040

On irradiation at 366 nm, the fulgide **37e** in chloroform did not show any competing photochemical E → Z photoisomerization compared with the fulgide **37a** which had a quantum yield of E → Z photoisomerization of 0.13. The elimination of this undesirable side reaction may have significance in many applications for fulgides.

3.3.2.2 Alkylidene and arylidene substituents. The arylidene and alkylidene-substituted fulgides **39**, (R = alkyl or aryl) are both photochromic and thermochromic (Whittall, 1979; Heller *et al.*, 1981).

39 R=Alkyl, Aryl

When furyl fulgides **39** are synthesized via the Stobbe condensation method (see section 3.1) the (E,E) isomers are normally obtained. These fulgides are both photochromic (colouring to the *trans*-7,7a-DHBFs **41** by the normal conrotatory mode) and thermochromic. On heating, the over-crowded (E,E)-fulgides undergo thermal isomerizations to the less sterically hindered (E,Z)-fulgides **40** which then cyclize to the *trans*-7,7a-DHBFs in a disrotatory mode.

The red–purple 7,7a-DHBFs **41** can be bleached back to the (E,E)-fulgides **39** using visible light. It has been shown that the (E,Z)-fulgide **40** (R = Me) also undergoes conrotatory photocyclization to the *cis*-DHBF **42** (R = Me). This is bleached back to the fulgide **40** (R = Me) with white light (quantum yield of 0.11 at 500 nm, in toluene solutions at 23°C), which is nearly an order of magnitude larger than the quantum yield for the conversion of the *cis*-7,7a-DHBF **41** (R = Me) into the (E,E)-fulgide **39** (R = Me) (quantum yield of 0.013 at 500 nm in toluene solutions at 23°C) (Heller *et al.*, 1981).

3.3.2.3 Adamantylidene substituents. The furyl fulgide **35** had a relatively low quantum yield for the photochemical bleaching of the 7,7a-DHBF **36** back to the fulgide **35**. However, for reversible optical data storage technol-ogy, a high efficiency for photochemical bleaching is required as data recording rates depend on this reaction. Replacement of the isopropylidene group in the fulgide **35** by the adamantylidene group gives the fulgide **43a**, which had similar absorption spectra and fatigue resistance characteristics as the photochromic system based on the fulgide **35** but shows a sixfold increase in the quantum yield for bleaching (Whittall, 1979).

Heller (1989) reported that the fulgide **43b** had a high quantum yield for colouring (0.50 for 366-nm irradiation in toluene), while the corresponding

43   a R=Me b R=^{i}Pr   **44**   a R=Me b R=^{i}Pr

7,7a-DHBF **44b** had a high quantum yield for bleaching (0.26 for 546-nm irradiation at 26°C in toluene), which demonstrated that it is possible to make fulgide systems which have high quantum efficiencies for both colouring and bleaching reactions. This system was studied independently by Kurita *et al.* (1990a) with the same results. These authors also reported that the 7-norbornylidene fulgides **45a, b** gave the 7,7a-DHBFs **46a, b** which have quantum yields of bleaching which are similar to the 7,7a-DHBFs **38b–f** from the isopropylidene fulgides **37b–f** rather than from the adamantylidene fulgides **43a, b**.

45   a R=Me b R=^{i}Pr   **46**   a R=Me b R=^{i}Pr

3.3.2.4 Fluorenylidene substituents. The introduction of fluorenylidene groups into fulgides is known to introduce large bathochromic shifts of the longest wavelength absorption band which can extend the range of wavelengths for which the fulgides can be coloured. The deep-orange (Z)-fluorenylidene-substituted fulgide **48** in toluene was found to undergo Z → E photoisomerization on irradiation at 366 nm, but only a low conversion into the purple 7,7a-DHBF **50** was obtained. Heating a solution of the (Z)-fulgide **48** in orthodichlorobenzene at 180°C gave an initial reversible [1,5]-hydrogen shift to give the pale-yellow maleic anhydride **47** (after about

15 min), but prolonged heating (10 h) gave a quantitative conversion into the deep-purple 7,7a-DHBF **50** which could be efficiently bleached with white light to give the (E)-fulgide **49** (Whittall, 1979).

3.3.2.5 Modification of the furyl ring. The synthesis of photochromic furyl fulgides via the Stobbe condensation (section 3.1) used 2-methyl-3-acyl furan derivatives. The 2-methyl substituent is essential to prevent fatigue reactions in the 7,7a-DHBF derivatives formed on photocyclization. 2,4-Dialkyl-3-acylfurans are too sterically hindered to undergo the Stobbe condensation efficiently, and hence only substituents on the 5 position of the furan ring in fulgides can be varied easily. Substituents in this position are in conjugation with the main chromophore of the 7,7a-DHBF, and hence variations allow a wide range of different coloured 7,7a-DHBFs to be made.

The fulgides **51a–d** had similar absorption characteristics, with only slight bathochromic shifts being observed for fulgides with electron-donating aryl groups in the 5 position of the furyl ring, as shown in Table 3.3. On the other hand, large differences were seen in the spectroscopic properties of the coloured forms, with the 7,7a-DHBF **52b** having a slight hypsochromic shift compared with the 7,7a-DHBF **52a**, while the 2-aryl-7,7a-DHBFs **52c, d** had large bathochromic shifts relative to 7,7a-DHBF **52a** (46 nm for **52d** in toluene solutions) and they also had large hyperchromic shifts, the molar

51a R=Me   **c** R=Ph
   b R=H   **d** R=p-MeO-Ph

52a R=Me **b** R=Ph
   c R=H **d** R=p-MeOPh

extinction coefficient being 8200 dm³/cm/mol for **52a** but 17 800 dm³/cm/mol for **52d**. These data are shown in Table 3.4. These furyl fulgides are stable enough to allow electrophilic substitution (Heller, 1989).

Table 3.3 Spectroscopic data for the fulgides **51a–d** (toluene solutions).

	Fulgide			
	51a	**51b**	**51c**	**51d**
λ_{max} (nm)	343	330	354	366
ε_{max} (dm³mol/cm)	6780	6920	6305	6000
Φ_{col}	0.20	0.19	0.19	0.22

Table 3.4 Spectroscopic data for the 7,7a-DHBFs **52a–d** (toluene solutions).

	7, 7a-DHBF			
	52a	**52b**	**52c**	**52d**
λ_{max} (nm)	494	473	520	540
ε_{max} (dm³mol/cm)	8200	6950	15750	18200
Φ_{rev}	0.06	0.10	0.014	0.004

The fulgide **35** reacts with acid chlorides in the presence of tin(IV) chloride, to give the fulgides **53** with an acyl group on the 4 position of the furyl ring (Hibino and Ando, 1987a,b). These undergo photochemical cycliz-ation on UV irradiation to give the 7,7a-DHBFs **54**, which have $\lambda_{max} \sim 480$ nm and hence show a hypsochromic shift of about 14 nm compared with the 7,7a-DHBF **36**. When unbranched long-chain acyl groups were intro-duced by this method, the fulgides had amphiphilic properties that allowed the construction of Langmuir–Blodgett membranes with good photochromic properties. No detailed reports of the spectroscopic properties (quantum yields, fatigue resistance, etc.) have yet been published for these acylfuryl fulgides.

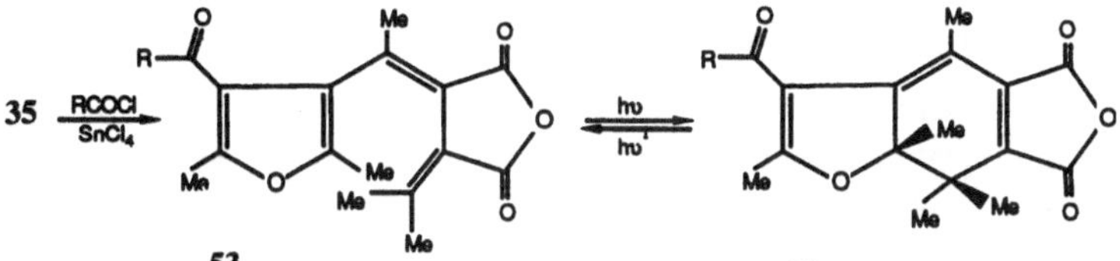

Fulgides **51b** may be nitrated by fuming nitric acid in acetic anhydride to give the nitrofuryl fulgides **55**, which photocyclize to the orange 7,7a-DHBF **56** on irradiation with UV light and can be bleached back to the fulgide **55** with white light (J. Whittall, unpublished results).

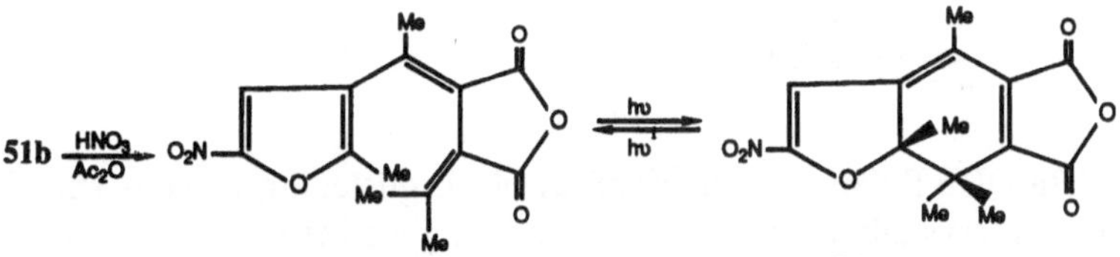

The photosensitivity of this fulgide is low, and there is an increase in the photochemical fatigue. This is thought to be because of the presence of lower energy, long-lived $n\pi^*$ triplet states associated with the nitro group which can increase photochemical side reactions.

A recent patent has also shown that the furyl fulgide **35** reacts with chlorosulphonic acid in pyridine to give water-soluble photochromic fulgides which are thought to have structure **57** (Trundle, 1987).

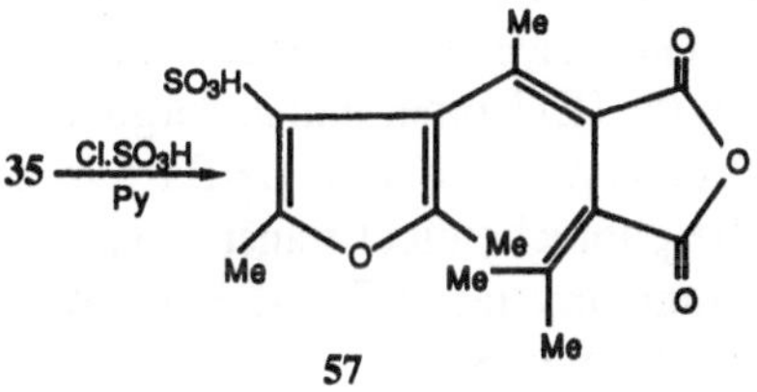

3.3.2.6 Other heteroaromatic substituted fulgides. Other heteroaromatic substituted fulgides have been made where the furyl ring has been replaced by benzofuryl, thienyl, benzothienyl, pyrrole, indolyl, oxazolyl, thiazolyl, isoxazolyl and pyrazolyl rings. A wide range of properties [colours, sensitivities (both colouring and bleaching), thermal and photochemical fatigue resistances and solubilities] have been observed.

(a) *Thienyl fulgides*. Thienyl fulgide (heteroatom exchange of sulphur for oxygen) has very similar properties to the furyl fulgide **35**. The main difference is a slight bathochromic shift of the absorption band of the 7,7a-dihydrobenzothiophene (7,7a-DHBT, **59a**) relative to the 7,7a-DHBF **36** (Heller *et al.*, 1985a).

58a R=Me b R=Ph   59a R=Me b R=Ph

The 7,7a-DHBT **59b** was obtained as black glistening crystals (the first example of a crystalline coloured form) and it gave the opportunity to compare the crystal structure of the fulgide **58b** with that of its 7,7a-DHBT **59b** (Heller, 1986).

Kaftory (1984) reported the X-ray crystal structures of the (E) and (Z) isomers of the fulgide **58b** and the corresponding 7,7a-DHBT **59b**. The diene system of the fulgide **58b** was skewed (torsion angles of 39° for the butadiene section and 15° for the anhydride section). The phenyl group was not coplanar with the thiophene ring in any of the compounds (twisted by 17° in the fulgide **58b** and 11° in the DHBT **59b**).

Single crystals of the (E)-fulgide **58b** changed colour upon irradiation, the external shape of the crystal was not damaged, and the unit cell dimensions of the (E)-fulgide **58b** were very similar to the unit cell dimensions of the 7,7a-DHBT **59b**. Consequently it was hoped that the photochromic transformation might take place in the single crystal. However, closer examination revealed that the (E) isomer had undergone spontaneous resolution on crystallization but the crystals of 7,7a-DHBT **59b** consisted of a racemic mixture. Therefore, the photochemical transformation of a single crystal of the fulgide **58b** to a single crystal of 7,7a-DHBT **59b** could not occur. The photocolouration of the fulgide **58b** was a surface effect.

(b) *Pyrrolyl fulgides*. The pyrrolyl fulgides **60** photocyclize very readily to give the 7,7a-dihydroindoles (DHIs) **61**, which have bathochromic shifts of more than 100 nm relative to the corresponding 7,7a-DHBFs (Heller, 1983).

The 7,7a-DHI **61b** shows a major solvatochromic effect (λ_{max} = 580 nm (ε = 7750) in hexane, 610 nm (ε = 7000) in toluene and 642 nm (ε = 9300)

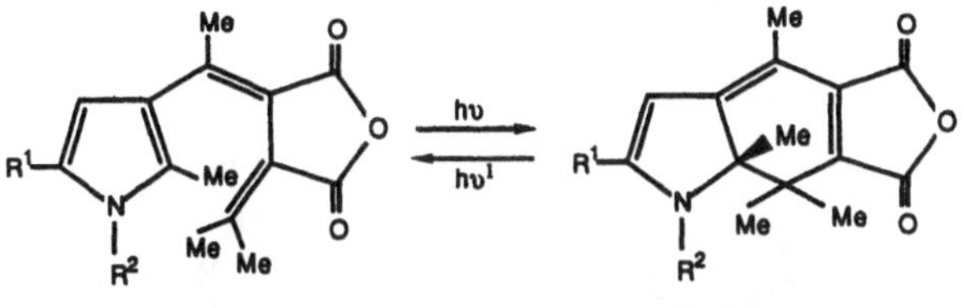

60a R^1=R^2=Me
  b R^1=Me R^{2,Ph}

61a R^1=R^2=Me
  b R^1=Me, R^2=Ph

in acetonitrile). The colour obtained is dependent on solvent polarity.

(c) *Benzofuryl and benzothienyl fulgides.* Benzofuryl and benzothienyl fulgides **62** (X = S, O) have been reported to have properties very similar to the corresponding furyl and thienyl fulgides (Heller, 1974).

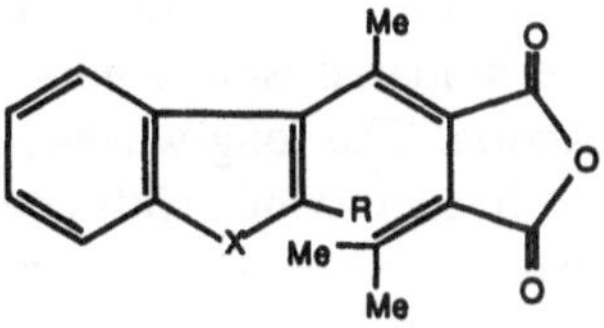

62 X=O,S

(d) *Indolyl fulgides.* The indolyl fulgides **63** have been studied by Russian (Minkin *et al.*, 1986; Grishin *et al.*, 1989), Japanese (Matsushima *et al.*, 1988) and Chinese (Wang *et al.*, 1989) research groups.

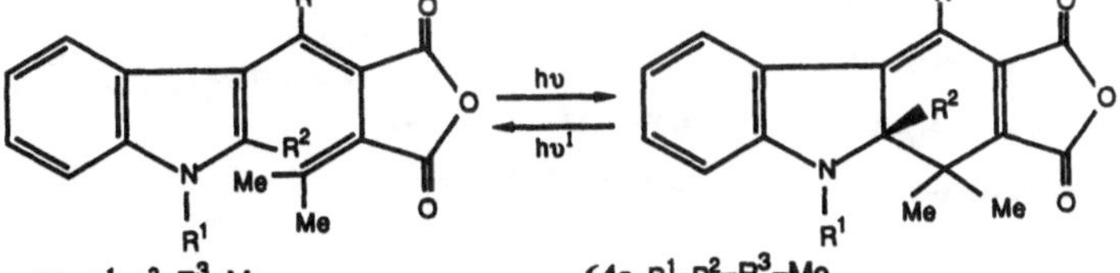

63a R^1=R^2=R^3=Me
  c R^1=Me, R^2=Ph, R^3=H

64a R^1=R^2=R^3=Me

The fulgide **63a** changes from pale-yellow to blue–green on irradiation with UV light. The X-ray structure of this photochromic indolyl fulgide was compared with that of the non-photochromic fulgide **63c** (Wang *et al.*, 1989) and it was noted that the conformations were completely different, with the two carbon atoms which form the new bond on photocyclization being very close together for the photochromic fulgide **63a** while the non-photochromic fulgide **63c** adopts a conformation in which these carbons are much further apart (Figure 3.10).

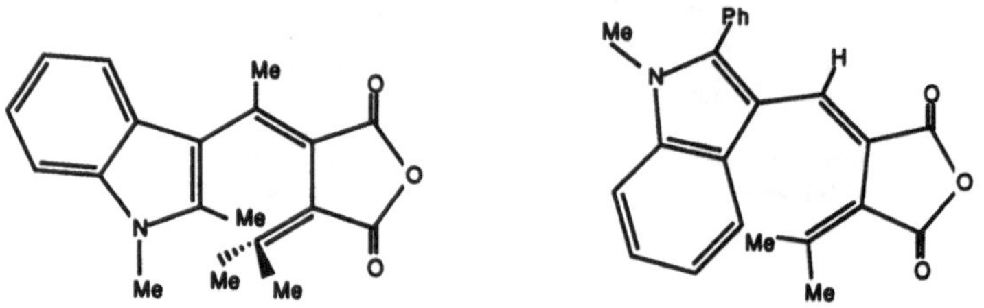

Figure 3.10 Conformations of the fulgides with structures 63a and c in solid state showing the *cis–cis–cis* conformation for the photochromic fulgide with structure 63a and the *cis–cis–trans* conformation for the non-photochromic fulgide with structure 63c.

(e) *Oxazolyl and thiazolyl fulgides.* The fulgides **65a–c** have been studied by Matsushima *et al.* (1988, 1989). The quantum yield for conversion of the fulgide **65a** in toluene into dihydrobenzoxazole (DHBO) **66a** at 366 nm is 84%, while the slightly more hindered isomer **65b** had a quantum yield of 91% for the same photoreaction. The long-wavelength absorption band of the DHBOs **66a, b** showed a hypsochromic shift compared with the corresponding 7,7a-DHBFs but were reported to have greater photochemical fatigue resistance, especially in thin polymer films.

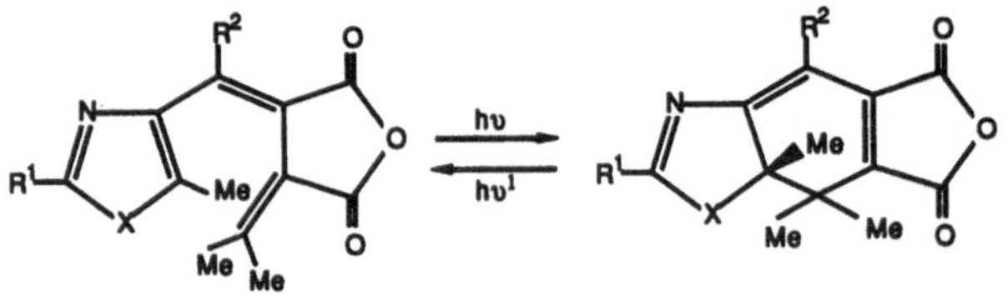

65a R^1=Ph, R^2=Me, X=O

b R^1=Ph, R^2= $^nC_{17}H_{35}$, X=O

c R^1=Ph, R^2=Me, X=S

66a R^1=Ph, R^2=Me, X=O

b R=Ph, R^2= $^nC_{17}H_{35}$X=O

c R^1=Ph, R^2=Me, X=S

84% conversion | 366 nm photo-
91% conversion | stationary state
for 0.2M $CDCL_3$ s

(f) *Pyrazolyl and isoxazolyl fulgides.* These fulgides (**67a–c**) have been described by Matsushima *et al.* (1988) but have a poor photoresponse to colouring with UV light.

(g) *Fulgimides from heteroaromatic substituted fulgides.* The reaction of fulgides with primary amines or ammonia results in nucleophilic ring opening of the anhydride ring to give succinamic acids, which can then be cyclized to the succinimide derivatives which are called fulgimides (Heller *et al.*, 1968). These compounds have similar photochromic properties to the parent fulgides but now have further functionality which may allow uses

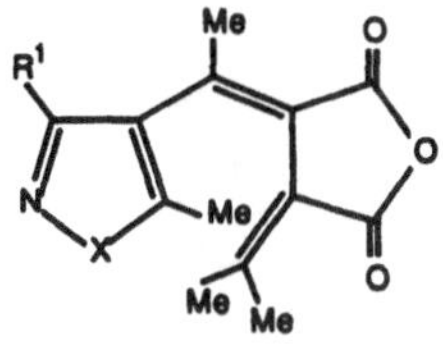

67a R^1=Me X=NPh
 b R^1=OMe X=NPh
 c R^1=Me X=O

which increase the potential for applications. This includes cross-linking with polymers (Smets and Deblauwe, 1988) and Langmuir–Blodgett film formation (Hibino and Ando, 1987).

3.4 Heliochromic compounds

3.4.1 Definition

A heliochromic compound was defined by Heller *et al.* (1985b) as a photochromic compound which colours in unfiltered sunlight and fades at ambient temperatures under diffuse daylight conditions or in the dark. By definition heliochromic compounds may be suitable for use in photochromic ophthalmic lenses. For this application compounds have to be near-colourless with a high efficiency for photochemical conversion to a highly coloured isomer which has a very low quantum efficiency for photochemical bleaching and a thermal bleach at ambient temperatures which is not too slow that the glasses do not clear sufficiently rapidly or too fast that they do not colour on a warm day. These compounds must be highly fatigue-resistant so that they are capable of retaining their photoreactivity for several years of use in sunglass application (see also Chapter 2).

3.4.2 Development of heliochromic fulgide derivatives

When the adamantylidene-substituted fulgide **68** was irradiated (366 nm) in organic solvents, the benzocyclobutene **72** was obtained and not the 1,8a-DHN **70** formed from the intermediate 1,2-DHN **69** as expected from analogy with the isopropylidene fulgides described earlier (section 3.2.3). However, heating the benzocyclobutene (100°C, 2 h in CDCl$_3$) yielded the 1,8a-DHN **70**, which on irradiation with 366-nm light regenerated the benzocyclobutene **72**. It was proposed that the *ortho*-quinodimethane **71** was an intermediate in this process (Whittall, 1979; Heller, 1983).

A flash photolysis study on a closely related compound revealed a short-lived intermediate ($t_{1/2} = 630 \pm 50$ μs at 23°C in hexane) which was assigned the structure **73**.

73

The benzannelated 1,8a-DHN **74** gave an orange *ortho*-quinodimethane **75** which could be observed by normal spectroscopic techniques and faded slowly to the benzocyclobutene **76** in toluene solutions at room temperature (J. Whittall, unpublished results).

Benzocyclobutene derivatives are well known while thienocyclobutene derivatives are rare. Heller *et al.* (1985b) synthesized the fulgides **77** where photocyclization onto the 2 position of a thienyl ring gave the 7,7a-DHBTs **78**, which underwent [1,5]-hydrogen shift to give the 6,7-dihydrobenzothiophenes (6,7-DHBTs, **79**). A range of these 6,7-DHBTs **79** were reported to give *ortho*-quinodimethanes on exposure to sunlight with a range of colours varying from orange–red to blue–green and with fade rates that gave half-lives ranging from a few minutes to several hours in organic solvents at room temperature.

Tanaka *et al.* (1987a,b, 1988) have synthesized many related compounds which have similar heliochromic properties. These include ones with various

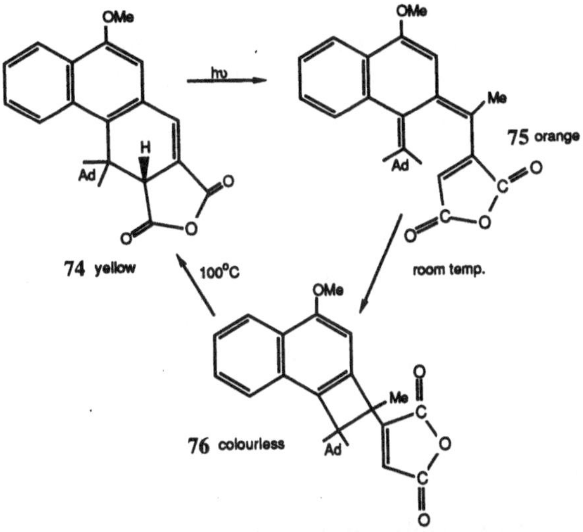
OMe
hν
Me
75 orange
Ad
H
74 yellow
100°C
room temp.
76 colourless

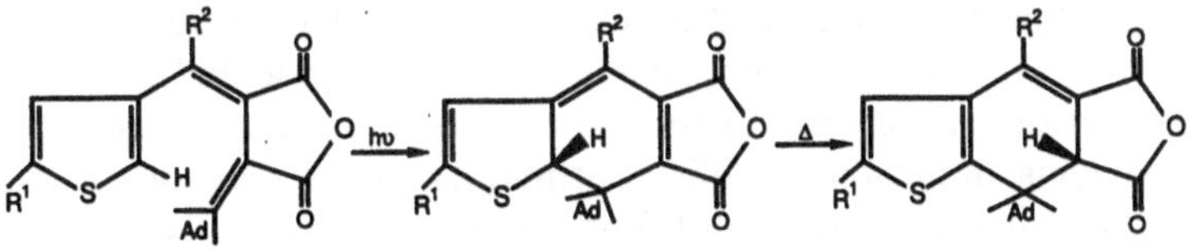
R²
R¹
hν
Δ
H
Ad

77 78 79

substituents on the adamantyl group and a range of fulgimides with a number of different substituents attached to the nitrogen atom. These workers demonstrated that compound **80** was also heliochromic.

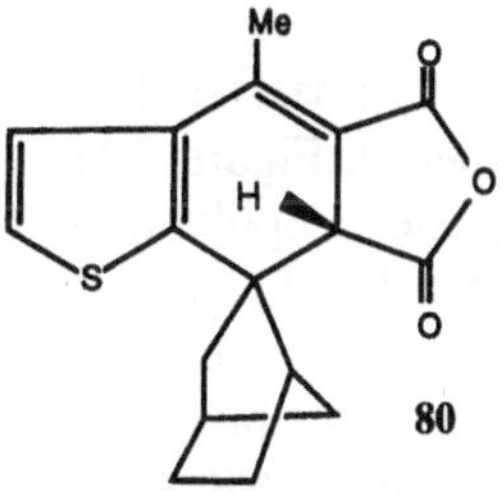
Me
H
S

80

The heliochromic compound **78** (R′ = H) could be methylated in the 6 position using methyl iodide in the presence of potassium carbonate to give the derivative **81**. When compound **81** was dissolved in poly(methyl methacrylate) the system was photochromic but the half-life for thermal decolouration had increased dramatically to 1200 h at 80°C.

3.5 Fulgides and fulgimides in polymer matrices

3.5.1 Introduction

For most technological applications photochromic compounds need to be incorporated into polymer matrices. Many techniques have been used to achieve this objective including spin-coating of fulgide and polymer from organic solvents, imbibition of fulgide into cross-linked polymers from organic solvents, injection-moulding techniques and curing of solutions of fulgides in suitable monomers. These compositions have then been investigated to elucidate differences in properties from those of the fulgides in solution and to determine whether fatigue resistance is high enough for practical applications. There have been several reports that suggest that the photochromic properties in polymer matrices are very similar to the properties found in organic solutions.

3.5.2 Investigations of photochemistry of fulgides in polymers

Smets and Deblauwe (1988) synthesized the (E) and (Z) isomers of the fulgimides **82** and **83** as shown in Figure 3.11.

The fulgimide (**83**)-styrene copolymer-(83-CP) containing 0.65% (mol/mol) **83** was prepared by heating **83** with styrene and 2,2'-azobutyronitrile in 2-butanone at 65°C for 12 h under an inert atmosphere. The copolymer was purified by precipitation with methanol. Incorporation of **83** was greater than 98%. Number- and weight-average molecular weights were determined by gel permeation chromatography to be M_n = 16 300 and M_w = 39 300. The glass transition temperature (T_g) was 67°C (as determined by differential scanning calorimetry). A second copolymer containing 0.53%

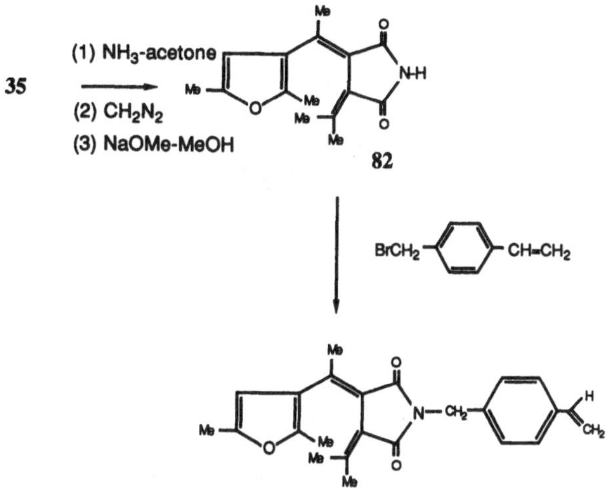

Figure 3.11 Synthesis of fulgimides with structures 82 and 83.

(mol/mol) of a 3:1 E–Z mixture of isomers of **83** [(E, Z)-83-CP] was also prepared with M_n = 12 600, M_W = 27 100 and T_g = 65°C. The spectroscopic properties of these copolymers were compared with those for (E) and (Z) isomers of the fulgide **35** and the fulgimide **82** both in solution in organic solvents and in polymer films containing 1.5–2.0% (w/w) of photochromic compound. Polymer films of approximately 35 μm thickness were prepared by casting on mercury surfaces from solutions in dichloromethane or chloroform followed by drying in vacuum at 25°C for 3 weeks. Cumene and methyl pivalate were used as solvents for the solution studies as these are good solvents for polystyrene and poly(alkyl methacrylates) respectively.

The quantum yields of the three main photoreactions (Z → E photoisomerization, colouring to the 7,7a-DHBFs and photobleaching of the 7,7a-DHBFs) were determined for the photochromic systems based on **35**, **82** and 83-CP both in solution in organic solvents and as cast polymer films.

In organic solvents the quantum yield for the photocolouration was independent of solvent, irradiation wavelength (in the 313–366 nm range) and temperature, consistent with the results of Heller and Langan (1981). In cumene solutions the quantum yields for photocolouration were 0.23 for **35**, 0.22 for **82** and 0.20 for 83-CP. Hence the attachment of fulgimides to polymer chains has very little effect on the photocolouring reaction in organic solutions.

The photochemical conversion of (Z) isomers of **35** and **82** into their (E) isomers was also determined. This reaction had quantum yields which were independent of wavelength of irradiation, was slightly dependent on solvent but was temperature-dependent. For example, the photoformation of **82** from its (Z) isomer in cumene using 316-nm radiation had a quantum yield of 0.206 at 21.5°C, which rose to 0.23 at 52.3°C and to 0.26 at 70.6°C. This is the first time that a temperature dependence for Z → E isomerization has been reported and warrants further investigation.

The quantum yields for photobleaching of the 7,7a-DHBFs back into the less coloured fulgide **35** and the fulgimide **82** using visible light were dependent on wavelength of irradiation, solvent and temperature, as discussed earlier (see section 3.3.1.1). The coloured 7,7a-DHBF from the 83-CP had quantum yields for photobleaching which differed only slightly from those for the non-bonded fulgimide analogues.

Quantum yields for these photoreactions were also determined for polymer films, which included **35** in polystyrene (PS) and poly(alkyl methacrylates) [alkyl = methyl, poly(methyl methacrylate) (PMMA); propyl, poly(propyl methacrylate) (PPMA); butyl, poly(butyl methacrylate) (PBMA)], **82** in PS and PBMA and for 83-CP. The increasing chain lengths of the alkyl substituents caused the T_gs to decrease from 91°C for PMMA, 26°C for PPMA and 12°C for PBMA for films containing **35**. The T_g for 83-CP was caused to drop from 67 to 32°C by addition of 0.8% dioctyl phthalate.

Above T_g the quantum yields for the colouring reactions were very similar to the solution values (i.e. **35** in PPMA at 50°C, Φ = 0.208; **82** in PBMA at 31.5°C, Φ = 0.204; and for 83-CP at 65°C, Φ = 0.182) and hence attachment to the polymer backbone had only a minor effect on the quantum yield for photocyclization. In all polymer matrices there is a small drop in quantum efficiency (~ 15–30%) in quantum yield below T_g (i.e **35** in PMMA or PS at 25°C, Φ = 0.166 or 0.158 respectively, for **82** in PS at 31.5°C Φ = 0.173 and for 83-CP Φ = 0.124 at 22.1°C), which means that the electrocyclization is affected to a small extent by the restriction of the mobility in these matrices below T_g.

The photobleaching reaction of the 7,7a-DHBF of the fulgide **35** behaves similarly, showing a slight increase in quantum yield to Φ = 0.091 in PPMA at 50°C compared with 0.075 in PMMA at the same temperature (with irradiation at 494 nm).

However, the most notable effect was a large difference in the quantum yield for the Z → E photoisomerizations for reactions above and below T_g. For the formation of **35** from its (Z) isomer at 50°C by irradiation at 343 nm the quantum yield was 0.064 in PS and 0.0436 in PMMA compared with 0.194 in PPMA. For the similar photoreaction to form **82** at 31.5°C with 316-nm irradiation Φ was 0.048 in PS but 0.169 in PBMA, and for photoconversion of (Z)-83-CP into (E)-83-CP (which had T_g at 39°C) with 316-nm light, Φ was 0.026 at 22°C and 0.167 at 65°C.

Hence, above T_g the quantum yields for Z → E photoisomerization were in the range 0.167–0.194, which is similar to the values found for solutions in organic solvents (0.20–0.28), while below T_g they were in the range

0.026–0.064. This is because of the large free volumes required to accommodate the major changes in conformation that occur during the Z → E isomerization.

Kurita *et al.* (1990b–d) studied 60–80 μm films containing about 1% (w/w) of the fulgides **35** and **37e** in polystyrene, poly(methyl methacrylate) and nitrocellulose by irradiation with 355-nm laser pulses (22 ps fwhm, 0.5–2 mJ output energy) at 22°C and then measured the rise time of the coloured 7,7a-DHBFs using picosecond resolution. For the 7,7a-DHBF **38e** the spectral band shapes at various delay times (4, 17 and 43 ps) were practically the same (which indicates no other coloured intermediates) and a rise time constant of 10 ps was observed in all three polymer matrices. Hence, this ultrafast formation of coloured 7,7a-DHBF takes place on the S_1 surface without an activation potential barrier, a result consistent with those obtained from time-resolved spectroscopic studies of fulgides in solution in organic solvents. That the photocyclization occurs prior to relaxation to the vibrational ground state of the first excited electronic state explains why the quantum yield for colouring is almost independent of environment, wavelength of irradiation and temperature.

However, the values for the quantum yields for both colouring and bleaching determined by these authors did not agree with the values determined by Smets and Deblauwe (1988) discussed earlier. Also, these authors concluded the glass transition temperature had no effect on quantum yields of E ↔ Z isomerizations. Further studies to clarify this situation would be helpful.

Ulrich *et al.* (1989, 1990) studied the thienyl fulgide **58b** as 0.1–0.01% (w/w) solutions in polystyrene platelets that were cast from dichloromethane solutions. The quantum yield for colouring **58b** to the 7,7a-DHBT **59b** using 366-nm irradiation was found to be 0.13 ± 0.05 for both dichloromethane solutions and in the polystyrene matrix and was independent of temperature. The quantum yield for bleaching of **59b** with 546-nm irradiation at 300 K was much higher (0.3 ± 0.1) in the polystyrene film than in dichloromethane solutions (0.01 ± 0.005). No explanation of this remarkable difference was proposed. This bleaching reaction was also thermally activated (activation energy 50 cm) and the quantum efficiency of ring opening dropped to < 0.001 at 4 K.

These authors also investigated photochromism in 0.1-μm films of the furyl fulgide **35** obtained by vapour deposition and found that the quantum yields for the colouring and bleaching reactions were similar to those found for solution studies.

One result that is not in complete agreement with the previous findings that the photochemical ring-opening reaction is thermally activated is the report (Horie *et al.*, 1988) that the 7,7a-DHBF **36** in poly(methyl methacrylate) films at 4 K had only 13% of the molecules in mobile sites; the 7,7a-DHBF was able to photoisomerize but for the molecules in these sites the quantum yield for ring opening at 514.5 nm was 0.04, almost the same as at room temperature (0.05).

Matsushima *et al.* (1988, 1989) studied a range of 14 heteroaromatic substituted fulgides in 1.0-μm poly(methyl methacrylate) films containing 15–20% w/w of fulgide and reported the relative photochromic properties and thermal and photochemical fatigue resistances. The photochromic properties of these fulgides were similar to the properties found in organic solvents, but the relative fatigue resistances were found to vary. For example the fulgide **35** retained its very good photochromic properties in these thin films, but these properties were reduced to less than 10% of their original value on heating the film at 80°C for 10 days. If the corresponding 7,7a-DHBF **36** was heated at 80°C, 90% of the colour was lost in a few hours. This system also underwent photochemical fatigue, and in 20 colour–erase cycles was reduced to half its initial concentration. By comparison, the thienyl fulgide **58a** had similar photochemical fatigue resistance, but the 7,7a-DHBT **59a** was much more thermally stable and only underwent 35% thermal degradation after 240 h at 80°C. The fulgides with the best balance of properties for technological application in these thin polymer films were the oxazolyl fulgides **65a, b** which gave a good photochromic response, underwent about 100 colour–erase cycles to fatigue to half the initial concentration and the coloured form of which underwent less than 10% degradation after heating at 80°C for 10 days. Addition of organonickel light stabilizers was found to improve the photochemical fatigue resistance by a factor of 2–3.

Dichroic films of the coloured 7,7a-DHBF **36** have been obtained (Jones *et al.*, 1989, 1990) by exposure of a poly(methyl methacrylate) film of the 7,7a-DHBF **36** to polarized light at 514.5 nm and an angular-dependent photoselection process results in anisotropy of the optical absorption which is stable unless the film is heated (to 50°C) when the molecular motions of the 7,7a-DHBF **36** molecules result in relaxation of the dichroism.

3.6 Applications of fulgides and fulgimides

3.6.1 Introduction

The potential applications that have been suggested for photochromic fulgides include actinometry, optical data storage, optical data processing and non-linear optical applications, optical waveguides, security and printing applications, eyewear products, contrast enhancement in photoresist technology and leisure products. The properties of organic fatigue-resistant photochromic fulgides have been compared with other photochromic compounds (Heller, 1983) and some of the applications of fulgides in polymer films for molecular electronics were discussed by Wilson (1984).

3.6.1.1 Actinometry. The use of the fulgide **35** as a UV actinometer (Heller and Langan, 1981) and the coloured 7,7a-DHBT **59b** as a visible actinometer (Heller, 1986) has been described in detail. These compounds offer the advantage over conventional actinometry methods of simplicity of use. The fulgide could also be used as an actinometer for the 308-nm excimer laser using optically dilute solutions by making allowances for the transmitted light (Scaiano *et al.*, 1988). The quantum yield for photoconversion of fulgide into the 7,7a-DHBF **36** was power-independent. It was also reported that the use of this actinometer in two-laser experiments involving the photolysis of short-lived reaction intermediates had the advantage that extinction coefficients of short-lived reaction intermediates could be calculated by a method involving fewer assumptions than methods previously used.

3.6.1.2 Optical data storage. The original suggestion (Hirshberg, 1956) that photochromic compounds could be used for rewriteable optical data storage systems was the impetus for the widespread current interest in photochromic systems. The requirements for a useful optical data storage system include a high storage capacity, no development of the recorded image, easy fabrication over large areas at low cost, high sensitivity for both recording and erasure, long archival lifetimes, and a system that can be recycled many times without loss of performance.

As the photochromism of fulgides is a molecular phenomenon it is a grainless system and hence high resolution (theoretically down to the molecular level) and full grey-scale recording are possible. The very nature of the photochromic response means no development of the recorded image is necessary, but this also means there is no gain (i.e. compared with the conventional photographic development process where there is chemical amplification of latent images). The kinetic studies of Kurita *et al.* (1990d) indicate that a writing speed of 10 GHz is possible when fulgides are used as optical data storage materials.

The fabrication of optical discs requires a support base (which can be flexible or rigid) onto which the photochromic active layer can be coated (possibly in a host matrix) and usually some form of encapsulation to protect the material from the environment and from physical damage. Fulgides have been incorporated into suitable matrices for optical data storage by diffusion into polymers, spin coating of solutions of polymer and fulgide using normal photoresist technology and by Langmuir–Blodgett film deposition. A fabrication method whereby the fulgide **35** was incorporated into a cross-linked polymer by coating a solution of the fulgide and benzoin ethyl ether in poly(ethyleneglycol) diacrylate as a 20-µm layer on a quartz substrate which was then cured by exposure to UV irradiation was described by Tsunoda and Suzuki (1987). This method gave a photochromic film which could undergo several hundred bleach and colour cycles without any significant degradation.

Ralston (1983) evaluated fulgides for optical data storage using polymer film solutions coated onto glass plates (0.5–10 μm thick) and found that suitable exposures for colouring were 5 J/cm^2 and that 300 mJ/cm^2 of 514.5-nm light caused an optical density change of 1.0 for bleaching. However, the plate was found to degrade on prolonged exposure to UV and was no longer active after the equivalent of 100 cycles. The environmental stability of the encapsulated plates was high.

Unlike other possible optical data storage methods (i.e. phase change or ablative methods) photochromic compounds do not have a threshold power for image degradation, and therefore if an absorption mechanism is used to detect the stored information the detection beam will cause some bleaching of the image. Two methods of overcoming this problem have been suggested: using a detection beam of a wavelength outside the absorption band of the coloured form of the photochromic compound allows the detection of either a phase change due to refractive index differences between the two states; or, using anisotropic media birefringence.

Wilson (1984) showed that the refractive index difference available for fulgide solutions in polymer matrices could be detected well into the infrared region of the electromagnetic spectrum. As phase changes are detected, this makes the system ideal for holographic data storage. Details of a holographic data storage system using the fulgide **43a** in polymer films were described by Heller (1979).

The detection of linear birefringence using anisotropic media (liquid crystal or stretched polymer film) has been described by Hattori *et al.* (1989). A few per cent by weight of the fulgide **35** dissolved in a nematic liquid crystal (Merck ZL1 1132) was placed in a 10-μm cell and aligned by rubbing in one direction, or a poly(vinyl alcohol) film was allowed to imbibe the fulgide **35** from ethanol and then the film was slowly stretched in hot air to 2–4 times its original length. These anisotropic media were then birefringent (i.e. linear polarized light is converted to elliptical polarized light by transmission through the sample). The photochromic reaction caused a change in birefringence which could be detected at 700 nm, which is well outside the absorption band of the 7,7a-DHBF **36**.

Similar results for the fulgide **43a** dispersed in a solution of the cholesteric polymer liquid crystal poly(δ-benzyl L-glutamate) as 2.5-μm films on a glass plate have been reported as a possible optical data storage medium (Suzuki *et al.*, 1986) which has a non-destructive read-out mechanism.

Another method of optical data storage that has been suggested is the use of the thermochromic fulgides **39** (section 3.3.2.2) in thin polymer films. By using an infrared laser to cause colouration in a thermal mode by a local heating mechanism, visible lasers could be used for read-out and erasure modes (Trundle, 1987). Several aspects of photochromic optical stores in practical system design have been described in Chapter 1.

3.6.1.3 Optical bistability, non-linear optics and optical data processing.
Optical bistability is the ability of a system to exhibit two distinct stable
optical transmission states for the same input intensity and has attracted
considerable attention for optical processing of data. The Fabry–Perot inter-
ferometer has two semireflecting plates, and when a material which exhibits
a non-linear absorption or dispersion is placed in between these optical bistability is often observed. When filled with a saturable absorber (i.e. a
compound for which the absorption coefficient is a decreasing function of
local light intensity), it exhibits hysteresis and bistability. Mitsuhashi (1981)
first demonstrated that toluene solutions of the fulgide **35** could be used as
a saturable absorber for green light (514.5 nm) in a Fabry–Perot filter and
had a critical power density of around 7 mW/nm^2, at which point transmission rose from 20 to 80% of input light. This method used a solution of
the coloured 7,7a-DHBF **36** in a 10^{-2} M toluene solution which was circulated
through the Fabry–Perot filter.

In the conventional saturable absorber, a stationary population of excited
and absorbing species is maintained by equilibrium between absorption and
decay processes, with the net result being an irradiation-dependent optical
absorption.

Kirkby *et al.* (1985, 1986) used 7.5-μm films of 1.5 M solid solutions of
the fulgides **35** and **43a** in poly(methyl methacrylate) in the cavity of a
Fabry–Perot filter and irradiated simultaneously with UV light (366 nm) at
a constant level of 2 mW/mm^2 and with visible light (514.5 nm) of variable
power. The transmittance shows hysteresis and bistability associated with
an absorptive/dispersive mechanism.

Fulgides can be incorporated into optical-grade plastic substrates by a
solvent-assisted diffusion process (Goodwin *et al.*, 1986). When 100-μm
films of poly[diethylene glycol bis-(allyl carbonate)] were heated with
perfluorinated hydrocarbon solution of fulgides at 165°C for 40 min the concentration profile of imbibed fulgide followed an exponential function with
a 1/e depth of 23 μm (Cush *et al.*, 1987). These samples had large intensity-dependent refractive index non-linearity (absent in solution deposited
layers) which had characteristics consistent with a thermally induced dispersive mechanism. In a Fabry–Perot filter, these samples then gave an optical
bistability when irradiated at 514.5 nm with a switch-up threshold intensity
of 2.33 kW/cm^2 and a switch-down intensity at 800 W/cm^2. The switching
time was ~900 μs. This technology could be used to produce arrays of compact devices which could be used as photochromic logic elements. Further
details have been covered in Chapter 1.

3.6.1.4 Optical waveguides. The transmission of signals by light guided
by fibre optics is now well established. Light is guided in a core region
which has a refractive index higher than that of the surrounding cladding
layer, which results in total internal reflection. Bennion *et al.* (1983)

described the use of thin layers of photochromic fulgides in polymers in the construction of stripe waveguides. By focusing a beam of UV light onto the substrate, the coloured 7,7a-DHBF stripe could be used to guide light of longer wavelength than that absorbed, and hence erasure of the higher refractive index state was avoided. The best method of construction of the optically flat surfaces necessary to avoid losses as a result of scattering was by using thermoplastic polymer solutions of fulgides sandwiched between glass plates. This method can be used to construct wavelength multiplexers and demultiplexers, and routing components (see Chapter 1 for further details).

3.6.1.5 Photochromic inks and paints. The potential of latent photochromic images to act as security marks on documents and labels has led to the development of several ink and dye compositions containing fulgides. Trundle and Brettle (1988) used a photochromic ink containing the fulgide **35** for screen printing photochromic labels, while Hawkins and Bowyer (1988) devised a photochromic film containing fulgides, on which images were obtained by overexposure of some of the film by irradiation through a suitable mask. A cellulose acetate solution of the fulgide **35** was spun into fibres that were woven into a fabric which was used to manufacture photochromic labels (Wright, 1988). (See Styron and Allinikon (1979) for application of photochromic paints in non-destructive testing.)

3.6.1.6 Eyewear applications. For this application the heliochromic compounds described in section 3.4 have to be incorporated into optical-grade polymers. Methods that have been described in the patent literature to achieve this objective include the imbibition of the heliochromic compound into CR-39 lenses either from the melt or by using inert perfluorinated solvents (Heller *et al.*, 1985b), injection moulding of the fulgides in an acrylic moulding process at 260°C (Baskerville *et al.*, 1985) and forming polyurethane laminates by incorporating the heliochromic compound into a di-isocyanate–polyester diol mixture followed by polymerization (Ormsby and Maltman, 1987). Chapter 2 describes this area in detail for spiroxazine photochromes.

References

Abdel-Wahhab, S.M. and El-Assal, S. (1968) *J. Chem Soc. (C)*, 867.
Abdel-Wahhab, S.M. and El-Rayyes, N.R. (1971) *J. Chem. Soc. (C)*, 3171.
Baskerville, M.W., Maltman, R.S. and Oliver, S.N. (1985) Eur. Pat. Appl. EP 136837.
Bennion, I. Hallam, A.G. and Stewart, W.J. (1983) *The Radio and Electronic Engineer* **53**(9), 313.
Boeyens, J.C.A., Denner, L. and Perold, G.W. (1988a) *J. Chem Soc. Perkin Trans. II*, 1749.
Boeyens, J.C.A., Denner, L. and Perold, G.W. (1988b) *J. Chem Soc. Perkin Trans. II*, 1999.
Brunow, G. and Tylli, H. (1968) *Acta Chim. Scand.* 22, 590.

Chakraborty, D.P., Sleigh, T., Stevenson, R., Swoboda, G.A. and Weinatein, B. (1966) *J. Org Chem.* **31**, 3342.

Cohen, M.D., Kaufman, H.W., Sinnreich, D. and Schmidt, G.M.J. (1970) *J. Chem. Soc. (B)*, 184.

Cohen, M.D., Bercovici, T., Fischer, E. and Sinnreich, C. (1977) *J. Chem. Soc. Perkin Trans. II*, 3654.

Cush, R., Trundle, C. Kirkby, C.J.G., Bennion, I. (1987) *Electronic Letts.* **23(8)**, 419.

Daub, G. and Johnson, W.S. (1951) *Org. Reactions* **6**, Chp. 1.

Freudenberg, K. and Kempermann, T. (1957) *Ann.*, 602, 184.

Goldschmidt, S., Riedle, R. and Reichart, A. (1957) *Ann.* 604, 121.

Goodwin, M.J., Glenn, R. and Bennion, I. (1986) *Electronic Lett.* **22(15)**, 789.

Grishin, I.Y., Chunaev, M.Y., Prizhyalgovskaya, N.M. and Suvorov, N.N (1989) *Khim. Geterotsikl. Soedin., Chem. Abstr.* **112**, 7, 309x (1990).

Harper, S.H., Kemp, A.D. and Tannock, J. (1970) *J. Chem. Soc. (C)*, 626.

Hattori, Y., Yoshitake, J. and Yamanaka, T. (1989) In *Chemistry of Functional Dyes*, eds. Yoshida Z. and Kitao, T. Toyko, 326, Mita.

Hawkins, M. and Bowyer, A.G. (1988) Eur. Pat. Appl. EP279600.

Heller, H.G. (1974) Brit. Pat. 1464603.

Heller, H.G. (1979) UK Pat. Appl. G.B. 2002752A.

Heller, H.G. (1983) *IEE Proc. J.* **130(1)**, 209.

Heller, H.G. (1986) In *Fine Chemicals for the Electronics Industry*, ed. Bamfield, P. RSC Special Publication 60, London, 127.

Heller, H.G. (1989) In *Chemistry of Functional Dyes*, eds. Yoshida, Z. and Kitao, T. Tokyo, 267, Mita.

Heller, H.G. and Hart, R.J. (1972) *J. Chem. Soc. Perkin Trans. I*, 1321.

Heller, H.G. and Langan, J.R. (1981) *J. Chem. Soc. Perkin Trans. II*, 341.

Heller, H.G. and Megit, R.M. (1974) *J. Chem. Soc. Perkin Trans. I*, 923.

Heller, H.G. and Oliver, S.N. (1981) *J. Chem. Soc. Perkin Trans. I*, 197.

Heller, H.G. and Szewczyk, M. (1974) *J. Chem. Soc. Perkin Trans. I*, 1487.

Heller, H.G., Hart, R.J. and Salisbury, K. (1968) *J. Chem. Soc. Commun.*, 1627.

Heller, H.G., Darcy, P.J., Strydon, J. and Whittal, J. (1981) *J. Chem Soc. Perkin Trans. I*, 202.

Heller, H.G., Glaze, Harris, S.A., Johncock, W., Oliver, S.N., Strydon, P.J. and Whittall, J. (1985a) *J. Chem. Soc. Perkin Trans. I*, **957**, 10–85.

Heller, H.G., Oliver, S.N., Whittall, J., Johncock, W., Darcy P.J. and Trundle, C. (1985b) UK Pat. Appl. G.B. 2146327A.

Heller, H.G., Darcy, P.J., Patharakom, S., Piggot, R.D. and Whittall, J. (1986a) *J. Chem. Soc. Perkin Trans. I*, 315.

Heller, H.G., Crescente, O. and Patharakorn, S. (1986b) *J. Chem. Soc. Perkin Trans. I*, 1599.

Hibino, J. and Ando, E. (1986a) Jpn. Kokai Tokkyo Koho JP 62, 242, 677.

Hibino, J. and Ando, E. (1986b) Jpn. Kokai Tokkyo Koho JP 63, 101, 475.

Hibino, J. and Ando, E. (1987a) Jpn. Kokai Tokkyo Koho JP 63, 273, 687.

Hibino, J. and Ando, E. (1987b) Jpn. Kokai Tokkyo Koho JP 01, 149, 887.

Hibino, J. and Ando, E. (1987c) Jpn. Kokai Tokkyo Koho JP 01, 09, 282.

Hirshberg, Y. (1956) *J. Am. Chem. Soc.* **78**, 2304.

Horie, K., Hirae, K., Kenmochi, N. and Mita, I. (1988) *Makromol. Chem. Rapid. Commun.* 9, 267.

Huang, F., Jiang, L. and Wu, G. (1984) *Ganguang Kexue Yu Kuang Huaxue* **2**, 42; *Chem. Abstracts* (1986) **104**, 59, 261e.

Ilge, H.D. (1986) *J. Photochem.* **33**, 333.

Ilge, H.D. and Colditz, R. (1990) *J. Mol. Struct.* **213**, 39.

Ilge, H.D., Paetzold, R. and Radeglia, R. (1984a) *J. Prakt. Chemie* **326(2)**, 222.

Ilge, H.D. and Paetzold, R. (1984b) *J. Prakt. Chemie* **326(5)**, 705.

Ilge, H.D. and Paetzold, R. (1984c) *Z. Chem.* **24(2)**, 70.

Ilge, H.D. and Schutz, H. (1984) *J. Prakt. Chemie* **326(6)**, 863.

Ilge, H.D., Ch. Drawert, Suhnel, J. and Paetzold, R. (1984a) *J. Prakt. Chemie* **326(2)**, 233.

Ilge, H.D., Hobert, H. and Paetzold, R. (1984b) *Z. Chem.* **24(3)**, 97.

Ilge, H.D., Kaschke, M. and Khechinashvili, D. (1986) *J. Photochem.* **33**, 349.

Ilge, H.D., Suhnel, J., Khechinashvili, D. and Kaschke, M. (1987) *J. Photochem.* **38**, 189.
Johnson, W.S., Dunnigan, D.A. and McCloskey, A.L. (1950) *J. Am. Chem. Soc.* **72**, 514.
Jones, P., Darcy, P., Attard, G., Jones, W.J. and Williams, G. (1989) *Mol. Phys.* **67**, 1053.
Jones, P., Jones, W.J. and Williams, G. (1990) *J. Chem. Soc. Farad. Trans.* **86(16)**, 1013.
Kaftory, K. (1984) *Acta. Cryst.* **C40**, 1015.
Kaptein, R., Kemmink, J., Vuister, G.W., Boelens, R. and Dijkstra, K. (1986) *J. Am. Chem. Soc.* **108**, 5631.
Kirkby, C.J.G. and Bennion, I. (1986) *IEE Proc. J.* **133(1)**, 98.
Kirkby, C.J.G., Cush, R. and Bennion, I. (1985) *Optics Commun.* **56(4)**, 288.
Kirkby, C.J.G., Cush, R. and Bennion, I. (1986) *Springer Proc. Phys.* **8**, 165.
Kurita, Y., Yokoyama, Y., Goto, T., Inoue, T. and Yokoyama, M. (1988) *Chem. Lett.*, 1049.
Kurita, Y., Yokoyama, Y., Iwai, T., Kera, N. and Hitomi, I. (1990*a*) *Chem. Lett.*, 263.
Kurita, Y., Kurita, S., Yokoyama, Y., Miyasaka. and Mataga, N. (1990*b*) *Abstracts of the Xlll IUPAC Symposium of Photochemistry, Warwick UK*, 154.
Kurita, S., Kashiwagi, A., Kurita, Y., Miyasaka, H. and Mataga, N. (1990*c*) *Chem. Phys. Lett.* **171**, 553.
Kurita, Y., Ito, H., Yokoyama, Y. and Hayata, Y. (1990*d*) *Bull. Chem. Soc. Jpn.* **63**, 1607.
Lenoble, C. and Becker, R.S. (1986) *J. Phys. Chem.* **90**, 2651.
Matsushima, R., Kaneko, A., Tomoda, A., Ishizuka, M. and Suzuki, H. (1988) *Bull. Chem. Soc. Jpn.* **61(10)**, 3569.
Matsushima, R., Suzuki, H., Tomoda, A., Ishizuka, M., Kaneko, A. and Furui, M. (1989) *Bull. Chem. Soc. Jpn.* **62(12)**, 3968.
Minkin, V.I., Medyantseva, O., Lyashik, O.T., Metelitsa, A.V., Andreeva, I.M., Knyazhaskii, M.I. and Volbushko, N.V. (1986) *Khim. Geterotsikl. Soedin.* **11**, 1569; *Chem. Abstracts* (1987) **107**, 58, 780.
Mitsuhashi, Y. (1981) *Opt. Letts. (USA)* **6(3)**, 111.
Modro, T.A., Davidse, P.A., Dillen, J.L.M., Heyns, A.M.J. and van Rooyen, P.H. (1990) *J. Chem.* **68**, 741.
Ormsby, M.E. and Maltman, W. (1987) Eur. Pat. Appl. EP 294,056.
Ralston, L.M. (1983) *Proc. SPIE* **420**, 186.
Santiago, A. and Becker, R.S. (1968) *J. Am. Chem. Soc.* **90**, 3654.
Scaiano, J.C., Wintgens, V. and Johnston, L.J. (1968) *J. Am. Chem. Soc.* **110**, 511.
Smets, G. and Deblauwe, V. (1988) *Makromol. Chem.* **189**, 2503.
Stobbe, H. (1893) *Ber.* **26**, 2312.
Stobbe, H. (1911) *Ann.* **380**, 1.
Strydon, P.J. (1974) *PhD Thesis*, Aberystwyth.
Styron, J. and Allinikov, S. (1979) *Sampe* **24(1)**, 364.
Suzuki, Y., Ichimura, K. and Masako, S. (1986) Jpn. Kokai Tokkyo Koho JP 62, 209, 186.
Swoboda, G.A., Wang, K.T. and Weinstein, B. (1967) *J. Chem. Soc. (C)*, 161.
Tanaka, T., Imura, T. and Kida, Y. (1987) Jpn. Kokai Tokkyo Koho JP 0, 190, 286.
Tanaka, T., Imura, T. and Kida, Y. (1987) Eur. Pat. Appl. EP 316, 179.
Tanaka, T., Tanaka, K., Imura, T. and Kida, Y. (1988) Eur. Pat. Appl. EP 351, 112.
Trundle, C. (1987) UK Pat. Appl. GB 2, 209, 751.
Trundle, C. (1987) UK Pat. Appl. GB 2, 208, 271.
Trundle, C. and Brettle, J. (1988) PCT Int. Appl. WO 8, 801, 288.
Tsunoda, M. and Suzuki, Y. (1987) Jpn. Kokai Tokkyo Koho JP 0, 108, 092.
Ulrich, K., Port, H. and Bauerle, P. (1989) *Chem. Phys. Letts.* **155**, 437.
Ulrich, K. and Port, H. (1990) *J. Mol. Struct.* **218**, 45.
Wang, F., Wang, H.–Z., Li, Y.–Z., Fan, G.–Y., Cui, X.–S. and Liu, Z.–C. (1989) *Chimica Sinica English Edition* **4**, 349.
Whittall, J. (1979) *PhD Thesis*, Aberystwth.
Wilson, A.E.J. (1984) *Phys. Technol.* **15**, 232.
Wright, P. (1988) Eur. Pat. Appl. EP 328, 320.
Yoshioka, Y. and Irie, M. (1989) In *Chemistry of Functional Dyes*, eds. Yoshida, Z. and Kitao, T. Tokyo, 302, Mita.
Yoshioka, Y., Tanaka, T., Sawada, M. and Irie, M. (1989) *Chem. Letts.* 19.

4 Photochromic liquid crystal polymers

V. KRONGAUZ

4.1 Introduction

4.1.1 Mesomorphic phases

Thermotropic liquid crystals are ordered organic liquids characterized by orientation and anisotropic properties. The liquid crystals are formed on melting of crystalline solids and are converted to the isotropic liquid upon further heating, which is why liquid crystals are said to form a mesomorphic phase or mesophase. The temperature at which such a transition takes place is called the clearing temperature (T_c), because at this temperature the turbid mesophase becomes an optically clear isotropic fluid.

A compound (mesogen) which forms a mesophase is usually characterized by an anisometric shape of its molecules, for example rod-like or disc-like shaped molecules. Some examples of rod-like mesogenes are shown in Table 4.1. Usually they are composed of two or more benzene, cyclohexane or heterocyclic rings connected in a linear molecule directly or through a bridging group. End groups, such as alkyl, alkoxy or cyano, are also incorporated into *para* positions on the rings.

The thermotropic mesophases can be divided into two basic types: nematic and smectic (Figure 4.1). The first has only directional order while the latter has in addition some positional order, such as a layer structure. If the molecules constituting a nematic phase are optically active they can form a chiral or cholesteric mesophase whose director follows a helix. In fact, a nematic phase is a special case of a cholesteric phase with an infinite helix pitch.

The combination of molecular order and fluidity in a liquid crystal results in several remarkable macroscopic properties of the bulk material such as an anisotropic dielectric constant, birefringence, sensitivity to relatively weak external stimuli such as magnetic and electric field or boundary conditions. The ability to control the orientation of the liquid crystal axis by alignment of a mesophase in electric or magnetic fields is a source of many practical applications based on optical effects induced by these fields. Another factor

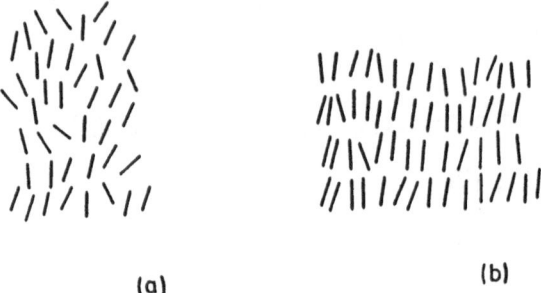

(a)   (b)

Figure 4.1 Schematic picture of the nematic (a) and smectic (b) phase.

Table 4.1 Selected examples of some basic structures of liquid crystalline compounds with rod-shaped molecular structure (Finkelmann, 1987).

of fundamental importance is the solid-surface boundary action, which tends to orient the long molecular axis of mesogens either parallel to the surface boundary (homogeneous alignment) or perpendicular to it (homeotropic alignment). A great deal of information on liquid crystals has been collected and published by Kelker and Hatz (1980). Among recent publications, those by Vertogen and De Jeu (1988) and Eidenschink (1989) can be mentioned; they contain more detailed references.

Incorporation of mesogenic groups into macromolecules may lead to liquid crystal polymers which yield a mesophase above the glass transition temperature (T_g) (Blumstein, 1985; Chapoy, 1985; Finkelman, 1987; Ringsdorf et al., 1988a; McArdle, 1989). In many cases the mesophase is frozen below T_g, resulting in ordered (mesomorphic) glasses. Two main types of liquid crystal polymers are recognized: (1) the main-chain (liquid crystal polymers with the mesogenic groups incorporated in the polymer chain) (Figure 4.2a, b) and (2) the side-chain (liquid crystal polymers with the mesogens attached to the main chain as side groups through a flexible spacer) (Figure 4.2c and d). Combinations of main- and side-chain liquid crystal polymers are also possible (an example is shown in Figure 4.2f). The liquid crystal polymers can form nematic, smectic and cholesteric mesophases similar to low-molar-mass liquid crystals.

4.1.2 Important photochromic systems

Spiropyrans and azoaryls as well as compounds related to these two types of photochromes are among the most comprehensively investigated

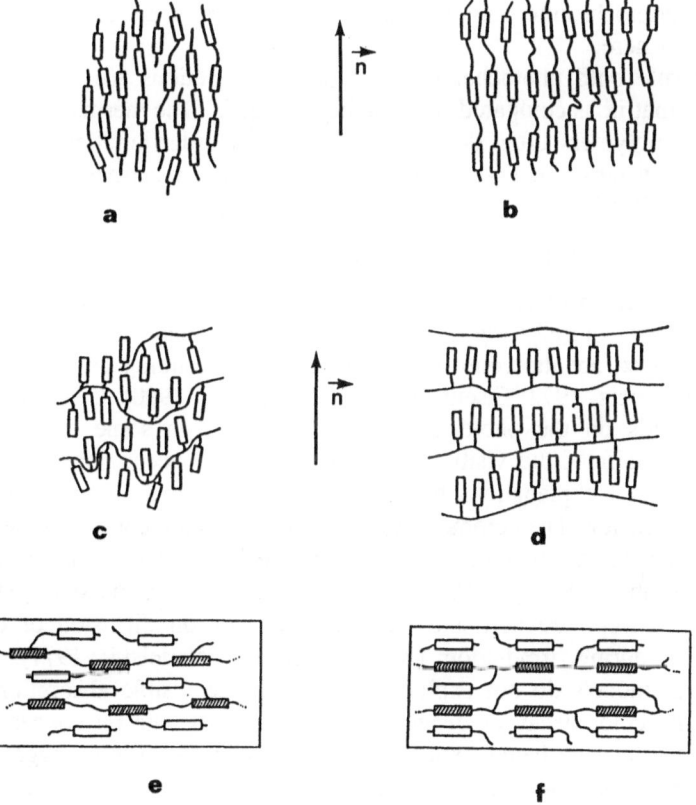

Figure 4.2 Schematic representation of main-chain liquid crystal polymers: nematic (a), smectic (b). Side-chain polymers: nematic (c), smectic (d). Combined main-chain/side-chain polymers: nematic (e), smectic (f). (After Vertogen and de Jeu, 1988; and Ringsdorf et al., 1988b.)

photochromic moieties which are incorporated in photochromic polymers and in liquid crystal polymers in particular (see Chapters 2 and 4 and Krongauz, 1990). It is important to emphasize here that the structural changes of photochromic molecules are accompanied by reversible changes of other properties of the molecules, such as refractive index, polarity, polarizability, etc. For some practical applications, changes in these properties may be more important than the color change (see Chapter 1).

The design of photochromic polymers is similar in a sense to the design of liquid crystal polymers: photochromes can be incorporated in a main chain or connected to a main chain (by a spacer) as a side group. In this chapter only the latter types with spiropyrans, spiroxazine and azobenzene side groups are discussed. The synthesis and structure of such polymers are described elsewhere (see also Kumar and Neckers, 1989; Krongauz, 1990). Some specific syntheses related mainly to spiropyran and azobenzene photochromic polymers are given in section 4.3.1. A few characteristic features of photochromic polymers important for further discussion should be mentioned here.

As a rule both photochemical and thermal reactions are retarded by a polymer matrix as compared with the liquid matrix. The thermal color decay in glassy polymers often deviates from the first-order kinetics and can be described in terms of two or even three first-order reactions with the rate constants differing often by an order of magnitude:

$$D = D_0[\alpha, \exp(-k_1 t) + \alpha_2 \exp(-k_2 t) + \alpha_3 \exp(-k_3 t)] \tag{1}$$

D and D_0 are, respectively, the current and initial optical densities of a photochrome colored form, k_1, k_2 and k_3 are the decay rate constants, and α_1, α_2 and α_3 are the contributions of the corresponding terms in the absorption (Smets, 1972; Eisenbach, 1978; Krongauz, 1990). Often the decay rate depends on the photochrome concentration. In some cases a hypsochromic shift of the visible absorption spectrum accompanies the color decay. Several explanations have been suggested by different authors for this phenomenon. The common hypothesis explains the complex kinetics by non-homogeneous distribution of free volume in a polymer matrix (Smets, 1972; Eisenbach, 1978). Other models are based on the dispersive first-order reaction concept (Richert and Bassler, 1985) or on diffusion of defects generated during irradiation (Kryszewski and Nadolski, 1977). These 'physical' models do not explain the shift of the visible absorption band during the color decay or dependence of the decay rate on concentration. These facts can be better explained in terms of photochrome aggregation, especially in the case of the polymers containing spiropyran pendant groups (Krongauz and Goldburt, 1981; Wismontski-Knittel and Krongauz, 1985). The colored merocyanine form of the photochrome has a very strong tendency to associate and to form molecular aggregates with stack-like structures (Parshutkin and Krongauz, 1974; Krongauz et al., 1978; Krongauz,

1979). The absorption spectra of the stacks are usually red-shifted compared with those of the isolated molecules, if the molecular dipoles are aligned in parallel (so-called J-aggregates). In the case of antiparallel dipole interactions the spectra are shifted to the blue (H-aggregates) (Sturmer and Heseltine, 1977). According to McRae and Kasha (1964), the spectral shift ($\Delta\rho$) is determined by the stack length according to the equation:

$$\Delta\rho \ (n \ \text{monomer} \ ) = 2(n - 1)<m^2>(1-3 \cos 2\alpha)/hnr^2 \qquad (2)$$

where h is Planck's constant, $<m^2>$ is the transition dipole moment of the monomer, r is the separation of molecular centers, α is the tilt angle between the line of centers and long molecular axes, and n is the degree of aggregation.

Estimation by eqn (2) indicated that even at a high concentration of spiropyran in a polymer, dimer and trimer H-stacks dominate the visible absorption spectrum of an irradiated film (Eckhardt et al., 1987). The spiropyran–merocyanine groups incorporated in side chains of macro-molecules are especially prone to aggregation. The H-type association of the merocyanine side groups was observed even in a dilute solution of the polymers (Goldburt et al., 1984a), while in very concentrated solution the aggregation (called 'zipper crystallization') proceeds up to the formation of three-dimensional polymeric crystals (Krongauz and Goldburt, 1981; Goldburt et al., 1984b; Wismontski-Knittel and Krongauz, 1985). The process appears on swelling of the polymer in the dark and is accompanied by spontaneous spiropyran–merocyanine conversion. The important feature of this process is cooperativity — the mutual stimulation of the chemical spiropyran–merocyanine reaction and the crystallization.

4.2 Search for photochromic mesophases — quasiliquid crystals (QLCs)

Thermotropic liquid crystals have acquired great importance for electro-optics because they change their optical properties in response to the application of an electric field. This allows conversion of an electrical signal into an optical signal as applied in liquid crystal displays.

Development of laser technologies has generated an interest in a mesophase which in addition would respond to light and could reversibly convert an optical signal into another optical signal. A possible approach to this problem would be synthesis of photochromic liquid crystals.

The most straightforward design of such a material would be a 'hybrid' molecule composed of mesogenic and photochromic units. Such combined molecules have been prepared by connecting photochromic spiropyran molecules with mesogenic groups (Table 4.2). The molecules shown in Table 4.2 can be classified as T-shaped and rod-like. In T-shaped molecules (com-

Table 4.2 Characteristics of spiropyrans containing mesogenic groups

No.	Substituent X	Y	Mesophase	Photochromism	Thermochromism
1	H	$CH_3O-Ph-COO-Ph-COO-(CH_2)_2-$	−	+	+
2	H	$CH_3O-Ph-COO-Ph-COO-(CH_2)_6-$	−	+	+
3	H	$CH-Ph-COO-Ph-COO-(CH_2)_6-$	−	+	+
4	$CH_3O-Ph-COC-C_6H_{10}-COO-$	CH_3	−	+	+
5	$CH_3O-Ph-COO-Ph-COO-$	CH_3	+	+	+
6	$C_5H_{15}-Ph-COO-Ph-COO-$	CH_3	+	−	+
7	$NC-Ph-COO-Ph-COO-$	CH_3	+	−	+
8	$NC-Ph-COO-Ph-COO(CH_2)_{11}-CONH-$	CH_3	+	−	+
9	$C_6H_{13}O-Ph-COO-Ph-CH=N-$	CH_3	−	+	+
10	$CH_3O-Ph-COO-Ph-CH=N-$	CH_3	−	+	+
11	$NC-Ph-COO-Ph-CH=N-$	CH_3	−	+	+

+, compound exhibits the indicated property.
−, compound does not exhibit the indicated property.

pounds **1–3**) a mesogenic group is attached to the nitrogen atom in the 1′ position through a flexible spacer, which is reminiscent of the structural principle of side-chain liquid crystal polymers. In rod-like molecules the mesogenic groups are connected to the photochromic moiety at the 5′ position either through a spacer (compounds **4** and **8**) or directly, forming a rigid molecular structure (compounds **5–7** and **9–11**).

The mesogenic substituents were composed of two rings connected by the bridging ester group and attached to the 5′ position of the spiropyran molecule by another bridging group (–COO–, –CH–N–, etc.). Synthetic routes for preparation of these combined compounds are shown in Figures 4.3, 4.4 and 4.5 (Cabrera *et al.*, 1984; Shvartsman and Krongauz, 1984a).

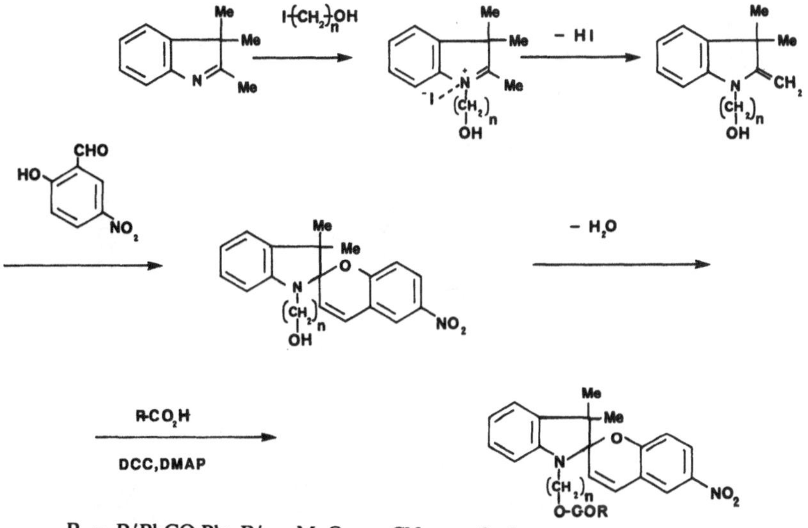

$R = R'PhCO_2Ph; R' = MeO –; –CN; n = 2, 6$

Figure 4.3 Synthesis of T-shaped spiropyrans (Zajtseva *et al.*, 1973).

The examples shown in Table 4.2 indicate that such a combination impairs either the mesomorphic or the photochromic properties characterizing the individual constituents, although thermochromism is preserved in all compounds.

The last three compounds listed in Table 4.2 form a metastable mesophase by heating their amorphous films, which can be obtained by casting from solution. The retardation of the film crystallization is caused apparently by the presence of merocyanine in low concentration which acts as an impurity. An interaction of mesophase with the glass surface seems to be an additional stabilizing factor. The alignment of the films in an electrostatic field also stabilizes the mesophase, which in supercooled conditions at room temperature is preserved practically indefinitely

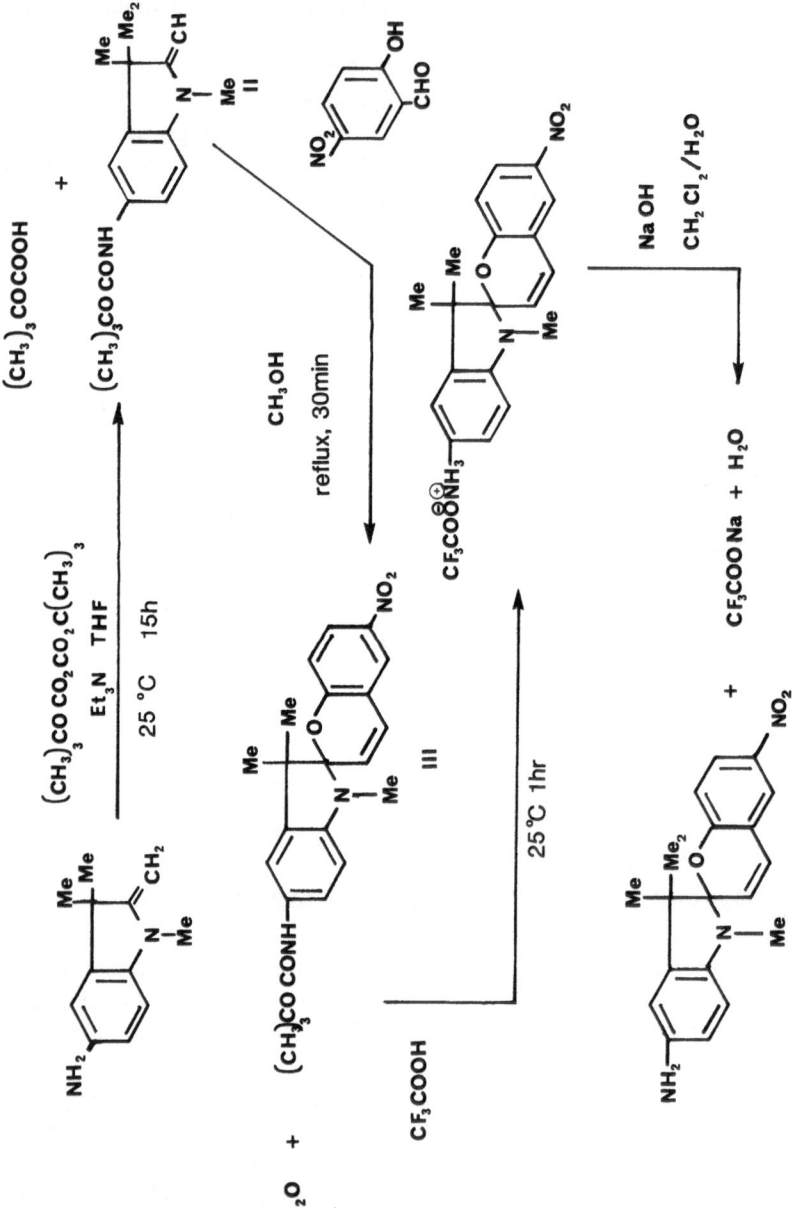

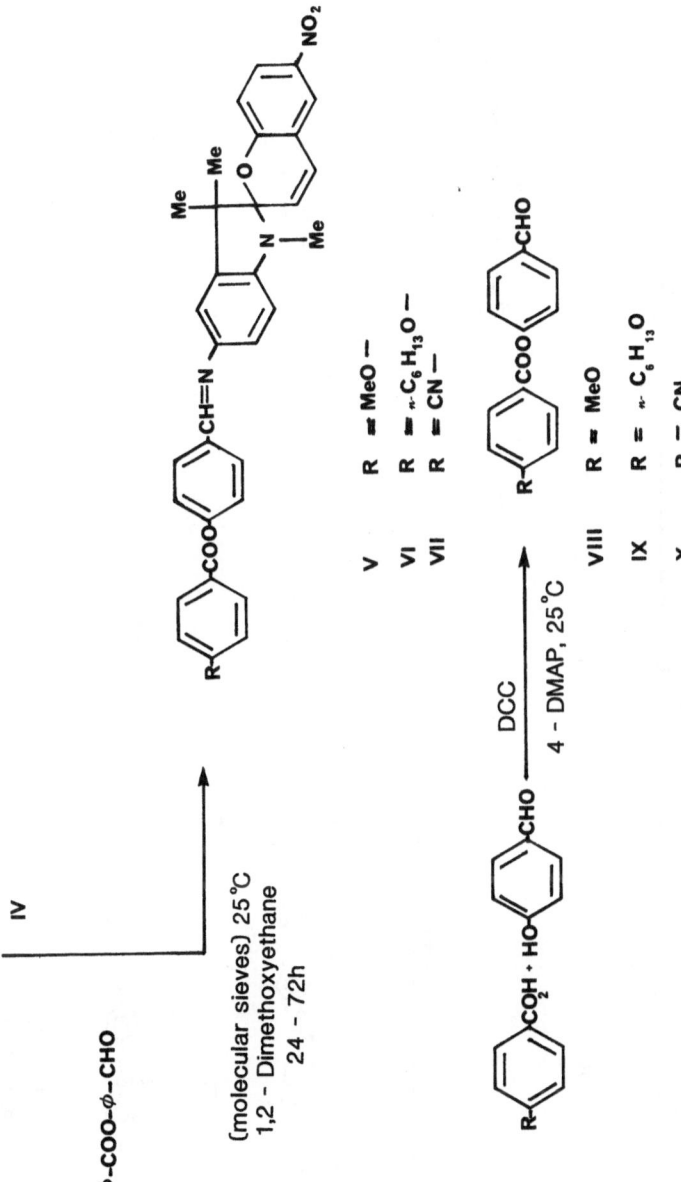

Figure 4.4 Synthesis of road-shaped spiropyrans (Shvartsman and Krongauz, 1989)

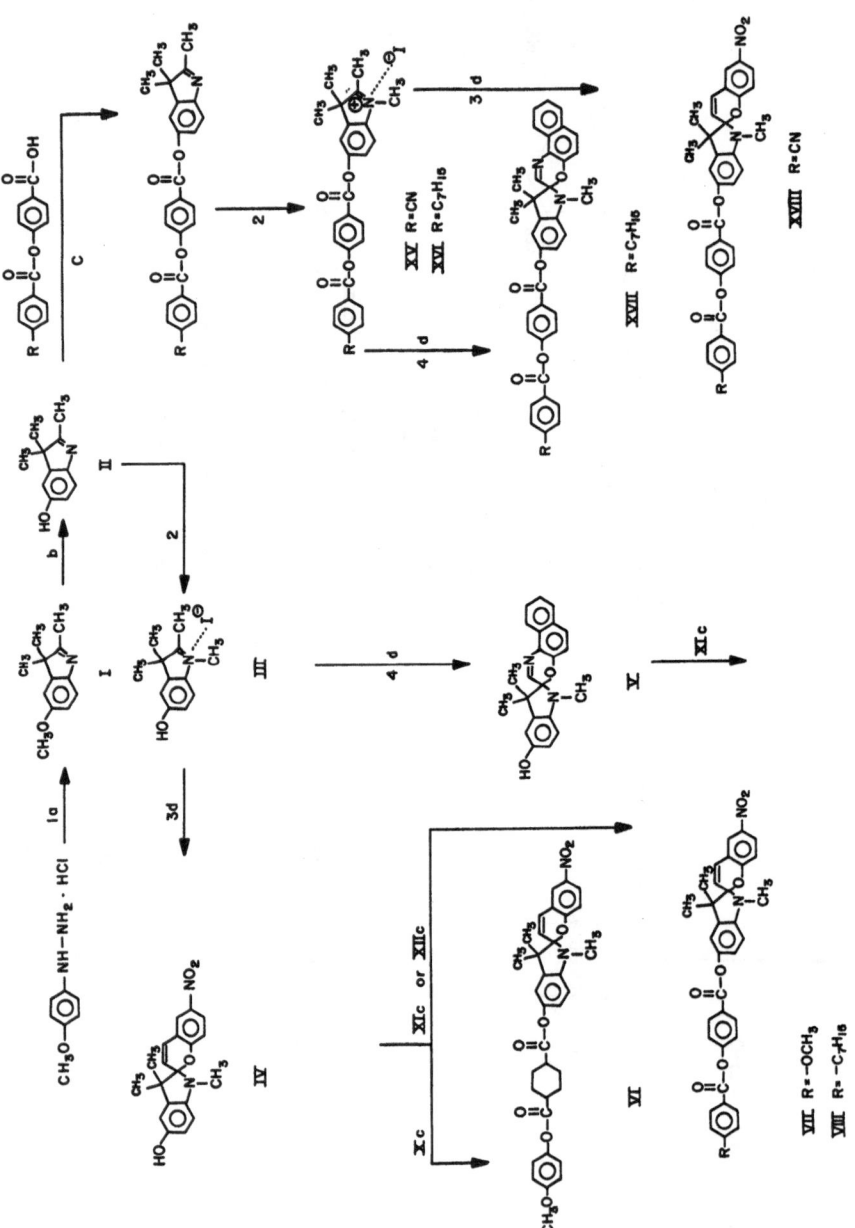

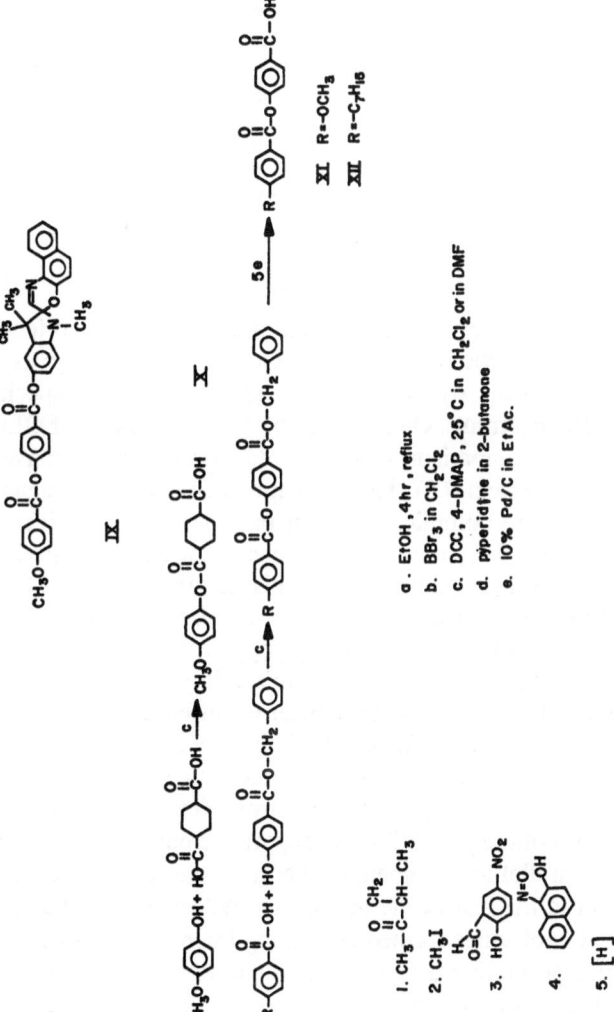

Figure 4.5 Reaction scheme for the synthesis of spiropyrans and naphthospiro-oxazines with mesogenic moieties connected by the ester bridging group (Shragina *et al.*, 1990).

(Shvartsman and Krongauz, 1984a, b). In this respect the mesophase resembles a liquid crystal polymer below T_g. These materials were named quasiliquid crystals (QLCs). The clearing points of QLCs lie much below the melting points of the corresponding crystals. The mesophase is not monotropic, i.e. it cannot be obtained by cooling of a crystal melt. This was explained by marked aggregation of merocyanine groups whose concentration in the melt is quite substantial.

The order parameters of the mesophase were estimated by different methods such as polarization of UV–visible and Fourier transform infrared (FT-IR) absorption spectra of oriented QLC films and electron spin resonance (ESR) measurements in the presence of a paramagnetic probe (Meirovitch et al., 1985; Shvartsman et al., 1985). In the optical measurements the order parameter (S) is defined as:

$$S = (D_{\parallel} - D_{\perp})/(2D_{\perp} + D_{\parallel}) \tag{3}$$

where $D_{\parallel}$ and $D_{\perp}$ are respectively the absorption of a chromophore oriented parallel and perpendicular to the mesophase director. This formula gives an order parameter of a dichroic dye (UV–visible spectrum) and a specific group in a molecule (IR spectrum). On average, the order parameter of QLCs is rather low (0.3–0.5) in comparison with the order parameters of conventional liquid crystals (more than 0.8). The estimation of the order parameter by the ESR method is more complicated but it provides the possibility of distinguishing a site where the paramagnetic probe is located. Comparison of the ESR and the optical measurements distinguishes two sites in the QLC structure: nematic domains embedded in a less organized amorphous-like site whose formation is promoted by bulky spiropyran units (Figure 4.6). X-ray diffraction of the QLC glass, studies of miscibility with other liquid crystals and measurements of the order parameter of different dichroic dyes doped to the QLCs have enabled refinement of this model. The QLCs represent an intrinsic two-component spiropyran–merocyanine mesophase with properties that are substantially determined by strong interactions between the molecules of these components. The benzopyran groups of spiropyran molecules distort the parallel arrangement of the molecules because of their non-planar attachment to the rest of the mesogenic molecules. On the other hand, the merocyanine molecules formed on heating improve the directional order in their vicinity, promoting a nematic mesophase. Obviously the two sites are in a thermal equilibrium which relates to the equilibrium between merocyanine and spiropyran molecules. The two-component nature of the mesophase is also a major factor in the stabilization of this metastable state which occurs much below the crystal melting point. At temperatures above T_c the aggregation of merocyanine is a factor which inhibits formation of the mesophase on cooling. Polar and highly polarizable merocyanine molecules are the source of non-linear optical properties of QLCs poled in an electrostatic field, as will be discussed later.

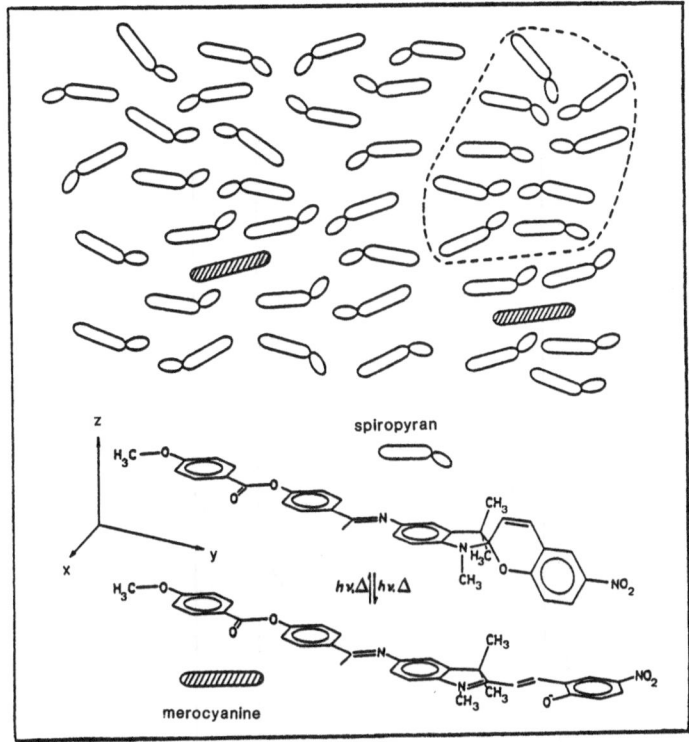

Figure 4.6 Structural model of the QLC mesophase. The area enclosed by the dotted line shows a possible amorphous site arrangement (Shvartsman *et al.*, 1985).

The merocyanine content of the QLCs can be varied only by changing the temperature because the materials are thermochromic but not photochromic. The absence of photochromic properties is associated presumably with attachment of a mesogenic substituent through a Schiff base bridging group to the 5' position of the spiropyran molecule, since the corresponding unsubstituted spiropyran or spiropyran substituted through the ester or amide bridging groups is photochromic. It is not yet understood whether this is connected with electron-donating properties of the Schiff base or with conjugation of the mesogenic group through the Schiff base.

Further efforts in searching for photochromic liquid crystals have been directed toward synthesis of spiroxazines containing mesogenic groups. Examples of such compounds are shown in Table 4.3, and the synthetic routes are given in Figure 4.5. Spiroindoline-naphthoxazine substituted with a mesogenic 4(4-heptylbenzoyloxy)-benzoloxy-group (structure **XVII** in Figure 4.5) proved to yield a mesophase with pronounced photochromic

Table 4.3 Characteristics of spironaphto-oxazines containing a mesogenic group.

No. at Figure 4.5	Substitutent X	T_c (°C)	T_m (m.p.) (°C)	Photochromism	Thermochromism
IX	CH_3O–Ph–COO–Ph–COO–	–	195	+	+
XVII	C_7H_{15}–Ph–COO–Ph–COO–	92	159	+	+

+, compound exhibits the indicated property.
−, compound does not exhibit the indicated property.

properties. Unlike QLCs, the mesophase is formed on cooling of the isotropic melt, but much like QLCs T_c lies far below the crystal melting point (see Table 4.3). Low ΔH peaks (~ 0.5 J/g) were observed both by differential scanning calorimetry (DSC) and by birefringent texture at the same temperature, both on cooling of the melt and on heating of the mesophase. The glassy birefringent films are stable at room temperature but prolonged heating of the films, especially above the clearing point-induced crystallization of the material.

The absorption spectra of the UV-irradiated spiroxazine XVII in the mesophase and in tetrahydrofuran solution are similar. In the visible region they have an absorption maximum at ~ 610 nm and a pronounced shoulder at ~ 580 nm that apparently indicates two overlapping bands (Figure 4.7). According to Schneider (1987) these bands relate to two merocyanine isomers which are in thermal equilibrium with each other. The fact that the shape of the spectra remains unchanged during the thermal color decay is in line with this assumption, rather than with the assumption of the aggregation of merocyanine (see section 4.1.2).

The color decay deviates from first-order kinetics and is relatively fast ($\tau_{1/2}$ ~ 5 × 10² s). The order parameter of the merocyanine dye is about 0.3, i.e. rather close to that of the QLCs.

Altogether these facts indicate that the structure of the spiroxazine mesophase is similar to the structure of spiropyran QLCs, although a distinguishing feature of the spiroxazine mesophase is photochromism. The author believes that the absence of an electron-accepting substituent in the naphthyl part of the molecule diminishes both the stability of the merocyanine form of the photochrome and its capacity to aggregate, resulting in a relatively short lifetime for the merocyanine.

The synthetic methods developed for preparation of spiropyrans and spiroxazines described in this section were used for synthesis of monomers employed for preparation of photochromic liquid crystal polymers described in the next section. Moreover, many features of QLCs and photochromic liquid crystal polymers resemble each other.

4.3 Photochromic liquid crystal polymers

Overall results described in the previous section indicate that the combination of photochromic and mesogenic groups in the same pendant moiety seldom gives photochromic liquid crystals and cannot be considered as a general way of preparation of such materials. The synthesis of side-chain liquid crystal copolymers containing both mesogenic and photochromic groups represents a much more universal approach (Figure 4.8). It opens many possibilities in molecular engineering because the mesogenic and photochromic groups are more autonomous in this case. For example, their

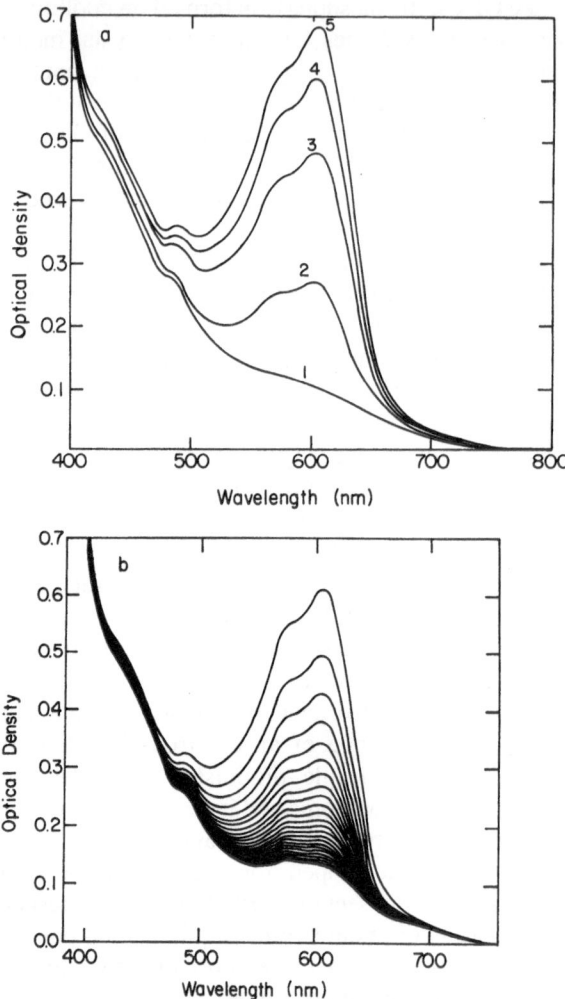

Figure 4.7 Absorption spectra of a glassy mesomorphic film of the photochromic spiro-oxazine at room temperature. (a) 1, before irradiation; 2–5, successive exposures to UV light ($\lambda = 365$ nm) for 1, 3, 5 and 10 min respectively. (b) Thermal decay of the visible absorption band measured at 2-min time intervals (Shragina *et al.*, 1990).

relative content in a macromolecule, the length of the flexible spacers, structure of the main chain, molecular weight and many other characteristics of a polymer can be tailored rather easily.

4.3.1 *Synthesis of copolymers*

Synthetic methods for the preparation of photochromic mesomorphic macro-molecules are basically similar to the methods for introducing other func-

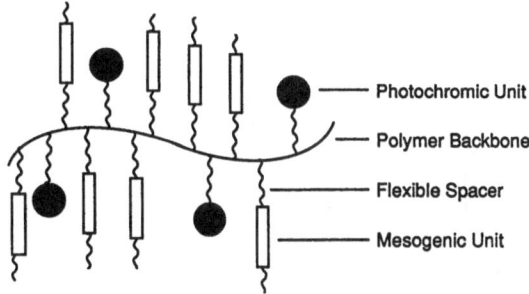

Figure 4.8 Schematic structure of a photochromic liquid crystal polymer.

tional groups into macromolecules. Usually copolymerization, condensation or polymer analogous reactions are used. Some photochromic monomers are also thermochromic and on heating generate a colored form which can be chemically very active (for example the merocyanine form appearing on heating of spiropyran solution). Therefore, special precautions should be taken to avoid excessive heating during reaction. Examples of synthetic routes for preparations of monomers and copolymers are given in Figures 4.9–4.12 (Portugal *et al.*, 1982; Ringsdorf and Schmidt, 1984; Cabrera *et al.*, 1987; Yitzchaik *et al.*, 1990a). One interesting feature of the polymer analogous reaction by which the spiropyran containing polysiloxane was prepared was synthesis of the intermediate active ester copolymers (Figure 4.12), because many functional groups on the spiropyrans interfere in the hydrosilation reaction. As a rule monomers comprising the copolymers differ in reactivity, therefore the composition of a copolymer differs from the composition of a monomer mixture. The molecular weight of acrylic and methacrylic spiropyran–mesogenic copolymers prepared by free radical polymerization was usually around 10^4 (Goldburt *et al.*, 1984b; Cabrera and Krongauz, 1987a) and did not depend markedly on the copolymer composition. In the case of polysiloxanes prepared by the polymer analogous reaction the starting polymer (Figure 4.12) contained about 35 monomer units.

$HO–C_6H_4–COOH + Cl–(CH_2)_n–OH \longrightarrow HO–(CH_2)_n–O–C_6H_4–COOH$

$H_2C=CH–COOH + HO–(CH_2)_n–O–C_6H_4–COOH \longrightarrow H_2C=CH–COO–(CH_2)_n–O–C_6H_4–COOH$

$HO–C_6H_4–NH_2 + O=CH–C_6H_4–CN \longrightarrow HO–C_6H_4–N=CH–C_6H_4–CN$

$H_2C=CH–COO–(CH_2)_n–O–C_6H_4–COOH + HO–C_6H_4–R$
$\xrightarrow{SOCl_2} H_2C=CH–COO–(CH_2)_n–O–C_6H_4–COO–C_6H_4–R$

$H_2C=CH–COCl + HO–C_6H_4–N=CH–C_6H_4–CN \longrightarrow H_2C=CH–COO–C_6H_4–N=CH–C_6H_4–CN$

$H_2C=CH–COCl + HO–C_6H_4–N=CH–C_6H_4–C \longrightarrow H_2C=CH–COO–C_6H_4–N=CH–C_6H_4–CN$

Figure 4.9 Synthesis of acrylic monomers with mesogenic units (Portugal *et al.*, 1982).

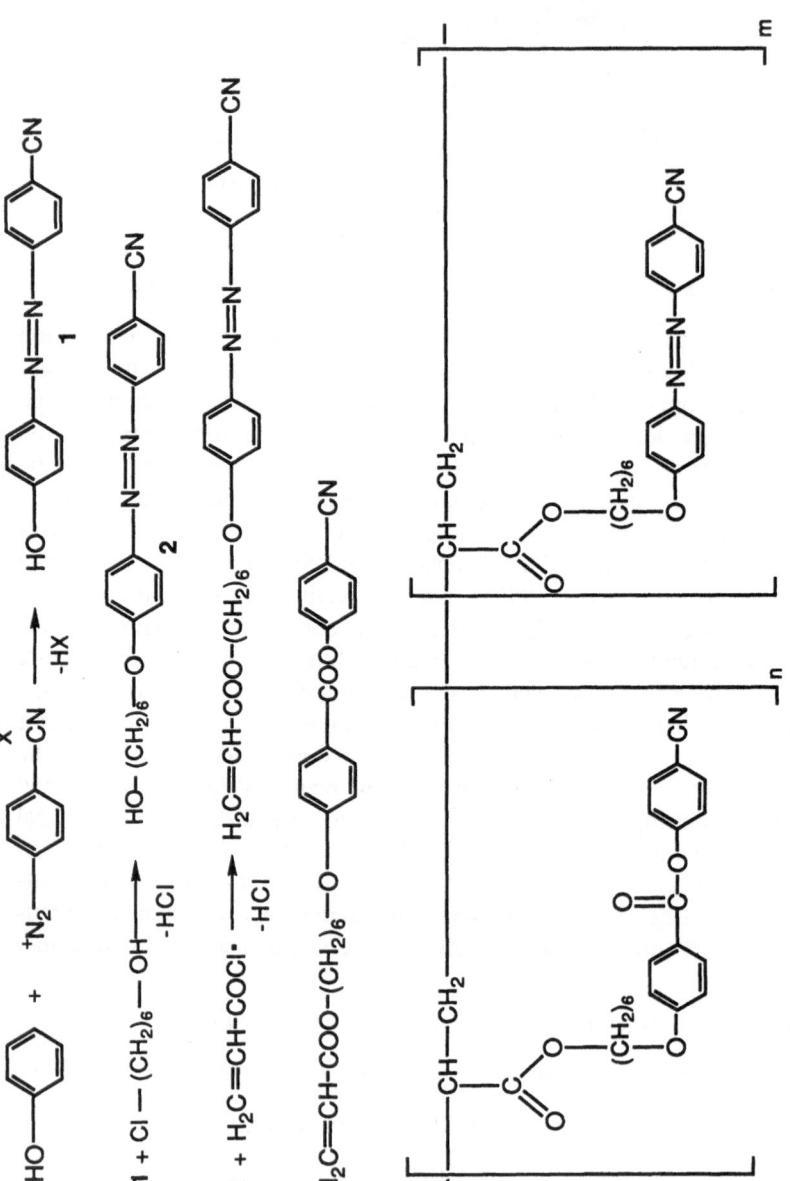

Figure 4.10 Synthesis of photochromic acrylic monomer and liquid crystal copolymers containing an azobenzene group (Ringsdorf and Schmidt, 1984).

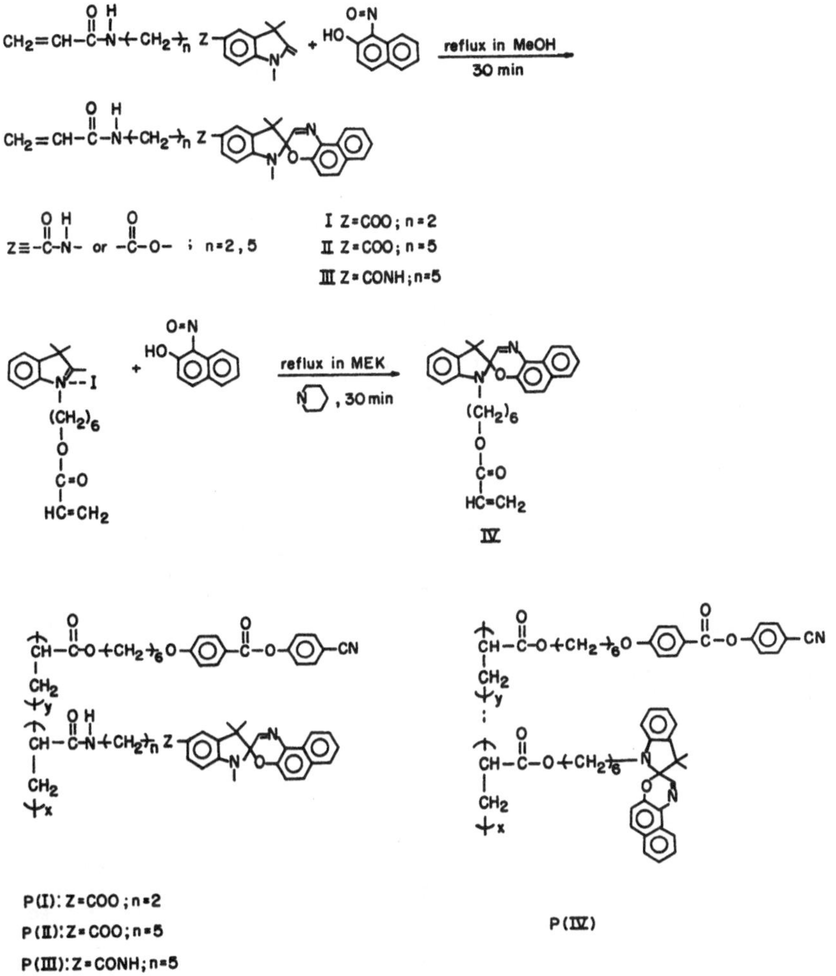

Figure 4.11 Synthesis of photochromic acrylic monomers and liquid crystal copolymers containing a spiroxazine group (Yitzchaik *et al.*, 1990b).

4.3.2 Thermodynamic properties

DSC measurements of phase transitions of photochromic liquid crystal copolymers can be inconclusive because the nematic–isotropic transition enthalpies are rather small for the homopolymers ($\sim$ 1 J/g). Incorporation of non-mesogenic photochromic units in the macromolecules leads to a further decrease of enthalpy and enhanced noise. The noise stems probably from the interactions of photochromic groups. Therefore the DSC measurements have to be combined with polarization microscopy (Cabrera and Krongauz, 1987b; Yitzchaik *et al.*, 1990b).

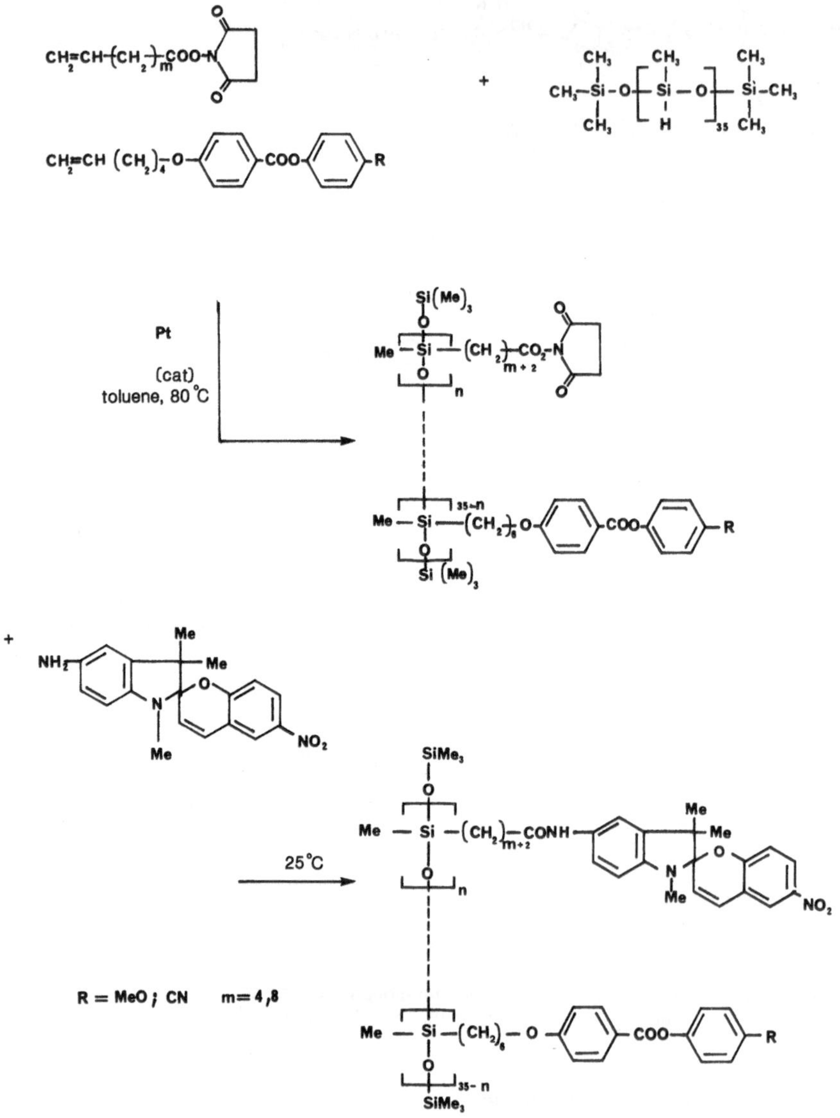

Figure 4.12 Synthesis of photochromic liquid crystal polysiloxanes containing spiropyran groups (Cabrera *et al.*, 1987).

In general, effects of a dye comonomer on the phase behaviour of a dye-containing liquid crystal polymer depends on the structure of the dye. The example shown in Figure 4.13 (Ringsdorf *et al.*, 1985) indicates that the introduction of bulky dye groups in a macromolecule narrows the tempera-ture range of the mesophase. The same is true for bulky photochromic dyes

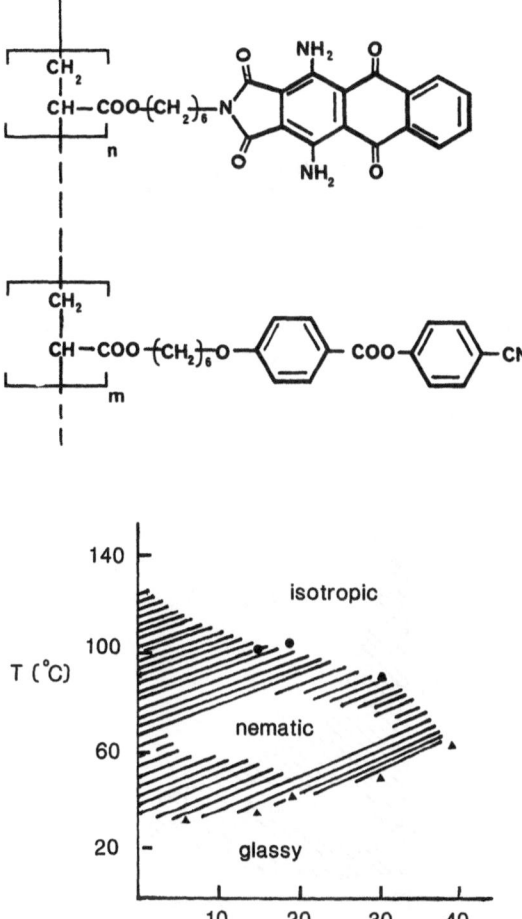

Figure 4.13 Phase diagram of the anthraquinone-containing copolymers (Ringsdorf *et al.*, 1985).

(Figure 4.14–4.16) (Cabrera *et al.*, 1987; Yitzchaik *et al.*, 1990b), while the rod-like azobenzene photochromes with distinct mesogenic character do not affect the mesophase range (Figure 4.17) (Ringsdorf *et al.*, 1985). The glass transition temperature (T_g) is not usually changed as markedly as the clearing point (T_c) with the increase of the photochrome content. The clearing point is also sensitive to the molecular weight of the copolymer. The copolymer with the higher molecular weight forms a more stable mesophase. Effects of molecular weight and polydispersity on thermodynamic characteristics were also observed for non-photochromic liquid crystal copolymers (Gray, 1989).

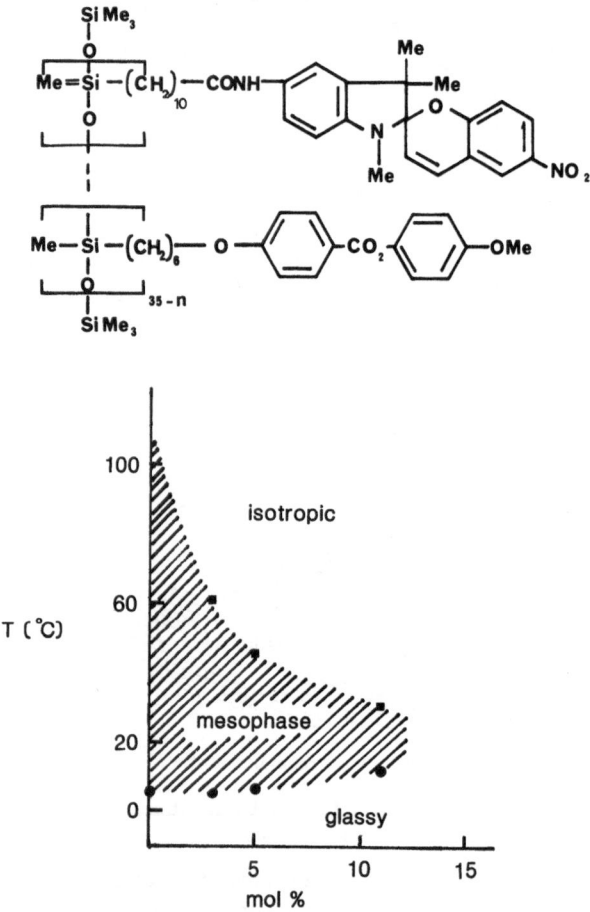

Figure 4.14 Phase diagram of the spiropyran-containing polysiloxanes. The composition of the copolymers is given as the percentage (mol/mol) of spiropyran groups (Cabrera *et al.*, 1987c).

Overall the mesophase formed by liquid crystal polymers is rather stable toward incorporation of bulky photochromic groups in a macromolecule. Even polymers with 40% of bulky dye units still reveal mesomorphic properties within a rather narrow temperature range.

4.3.3   Thermochromism

The thermochromism of the polymers containing spiropyran or spiroxazine

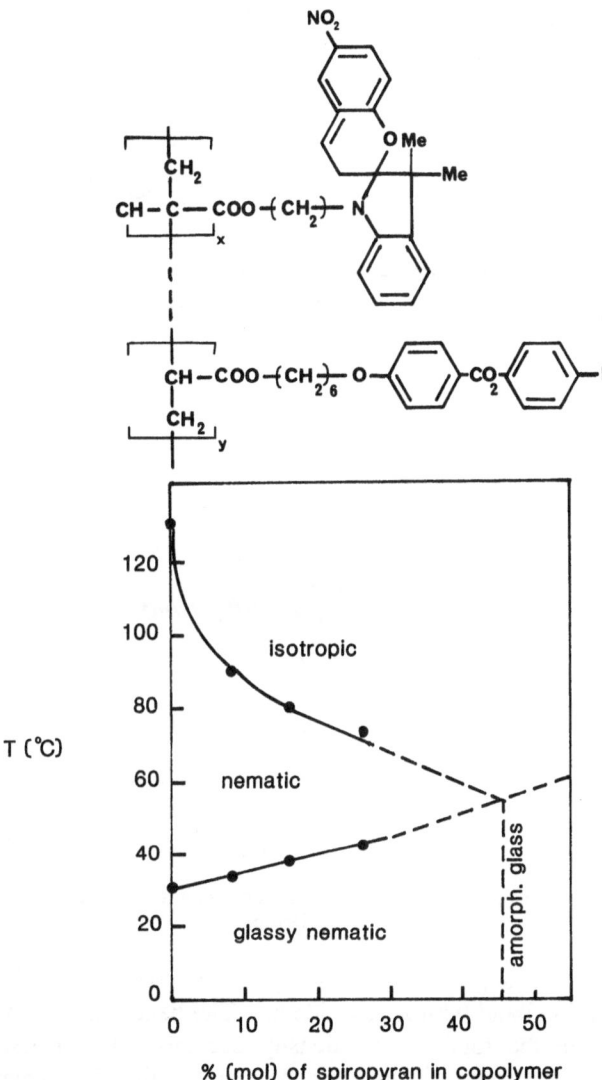

Figure 4.15 Phase diagram of the polyacrylates containing T-shaped spiropyran (Yitzchaik et al., 1990b).

side groups results from the spiropyran (spiroxazine) $\rightleftarrows$ merocyanine thermal equilibrium, which is shifted to the merocyanine on heating. The electronic absorption spectra of the polymer films show an increase of visible absorption with rising temperature (Figure 4.18) (Cabrera and Krongauz, 1987b). The transition to the liquid crystalline phase is accompanied by formation of a broad plateau in the range 460–585 nm. The optical density in this range does not change with temperature up to the clearing point. The

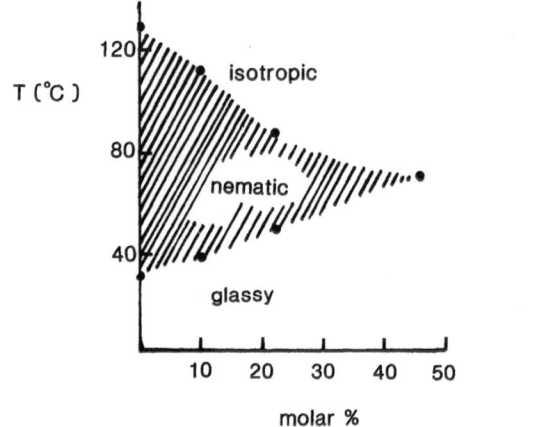

Figure 4.16 Phase diagram of the polyacrylates containing rod-shaped spiropyran (Yitzchaik *et al.*, 1990b).

transition from mesophase to isotropic phase coincides with the appearance of an absorption band with λ_{max} = 585 nm. The band vanishes reversibly on cooling, while the total optical density decreases. Comparison of these spectra with the spectra obtained for the merocyanine dyes aggregated as a result of swelling in a polar solvent (see also Goldburt *et al.*, 1984b; Eckhardt *et al.*, 1987) shows that the broad absorption in the range 460–585 nm is connected with association of merocyanine molecules in molecular stacks (probably very short ones, mostly dimers). The absorption of these associates overlaps apparently with the absorption of spiropyran (λ_{max} = 370 nm) and isolated merocyanine (λ_{max} = 585) side groups. The association of merocyanines should result in cross-linking of the macromolecules and, at high spiropyran content, in formation of a network, which is confirmed by elevation of the viscosity of the copolymer melt (see below). The conversion of the aggregated merocyanine groups back to spiropyrans on cooling of the

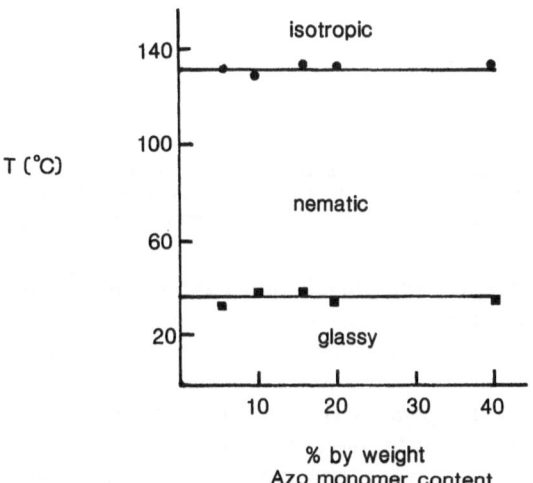

Figure 4.17 Phase diagram of the azo dye containing polyacrylates (Ringsdorf *et al.*, 1985).

copolymer is insignificant, and the color acquired by the films on heating is preserved after its cooling below T_g. The reversible absorption band with λ_{max} at 585 nm which characterizes non-aggregated merocyanine groups apparently becomes pronounced above the clearing point when formation of the network is completed and restricted segmental mobility hinders further merocyanine association.

4.3.4 Photochromism

4.3.4.1 Electronic absorption spectra. Irradiation with UV light of a glassy film formed after cooling below T_g of the mesomorphic copolymers containing spiropyran units produces a deep-blue color, λ_{max} in the range of 570–600 nm (Figure 4.19). The spectra belong to isolated merocyanine groups, though a blue shift of the spectra during the decoloration of a film

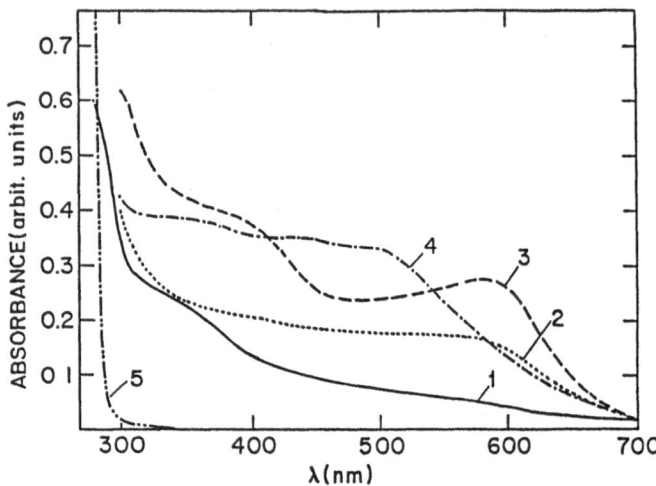

Figure 4.18 Absorption spectra of polyacrylates shown in Figure 4.16 (22% content of spiropyran): (1) amorphous film (26°C); (2) 82°C; (3) 95°C; (4) film swelled in tetrahydrofuran at room temperature; (5) homopolymer (no spiropyran).

revealed some aggregation (presumably dimerization) of the dye (Yitzchaik *et al.*, 1990b). The most spectacular color transformations occur in polysiloxane copolymers (Figure 4.20). The high flexibility of the chain segments, characteristic of polysiloxanes, leads to polymers with glass transitions below room temperature (Figure 4.14), which in turn dramatically affect the photochromic transformations in these polymers. The copolymer films cast from solution acquire a pink color at room temperature and form a mesophase (strong birefringence). Irradiation with visible light ($\lambda > 500$ nm) brings about a pale-yellow color, while irradiation of the yellow film with UV light ($\lambda = 365$ nm) results in a deep-red color. A yellow film irradiated with UV light at −20°C turns blue, and can be converted back to yellow by irradiation with visible light. The blue film is stable in the dark at −20°C but turns red upon heating above −10°C. The red film is thermally fairly stable but turns yellow on irradiation with visible light. At room temperature the fading of the red color takes several days or more, but less than a minute above the clearing point. The red color ($\lambda_{max} = 550$ nm) corresponds to aggregated merocyanines, which cause physical cross-linking of the macromolecules. The blue color ($\lambda_{max} = 580$ nm) has been ascribed to isolated merocyanine groups. The yellow color ($\lambda_{max} = 370$ nm) corresponds to the spiropyran absorption. The induced color transformation can be explained by the mechanism given in Figure 4.21. Apparently at temperatures below −10°C the side chains are immobilized (Horie *et al.*, 1985) and neither thermal dimerization of the merocyanine groups nor photochemical dissociation of their dimers occurs under these conditions. The possibilities of

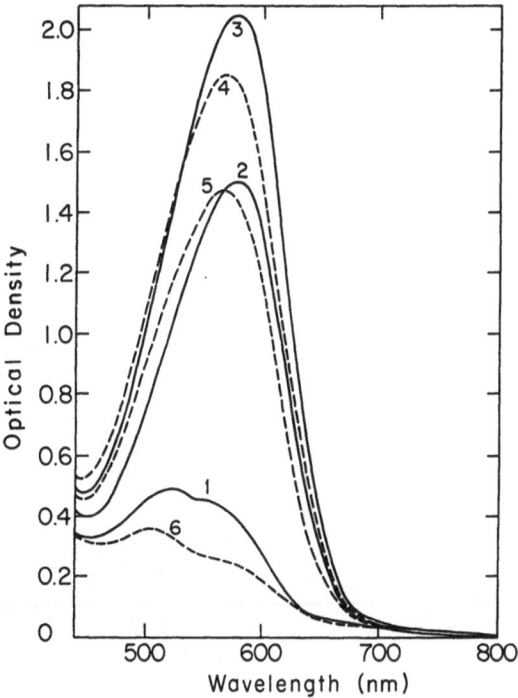

Figure 4.19 Absorption spectra of 21% copolymer depicted in Figure 4.16 (1) before irradiation; (2 and 3) after 12 and 24 min of successive UV irradiation; (4) after 17 h of thermal color decay at 25°C; (5 and 6) 3 min and 1 h irradiation with visible light (Yitzchaik *et al.*, 1990b).

controlling with light and temperature the formation of the three primary colors may make it possible to tailor these polysiloxane polymers for new applications in imaging technologies (Attard and Williams, 1987).

The copolymers containing spiroxazine photochromic units (Figure 4.11) have absorption maxima in the range of 610–630 nm and a pronounced shoulder around 580 nm, indicating two overlapping bands that were ascribed to two mercyanine isomers which are in thermal equilibrium with each other (Figure 4.22) (Yitzchaik *et al.*, 1990a). The shape of the spectra remains unchanged during the thermal color decay of the irradiated mesogenic glass, which is in agreement with the two-isomer assumption. Unlike polymers containing spiropyran units, no spectral indication of aggregation of mercyanines formed from spiroxazine was found.

The spectral changes characteristic of *trans–cis* isomerization of azobenzene in the liquid crystal polymers depicted in Figure 4.10 consist only in a decrease of the optical absorption without a marked shift of the absorption band, whose maximum remains at ~ 370 nm (Figure 4.23) (Wendorff and Eich, 1989). This means that only the changes of optical density without change in the color occur on UV irradiation of the polymer.

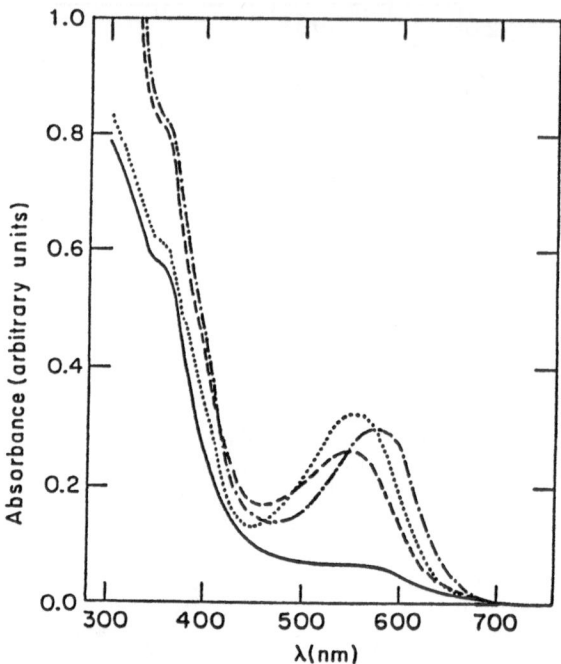

Figure 4.20 Absorption spectra of films of 11% copolymer depicted in Figure 4.14. —, yellow film at 25°C; •••, red film at 25°C; –•–, blue film at –20°C; – – –, red film obtained by heating the blue film to 25°C (Cabrera *et al.*, 1987).

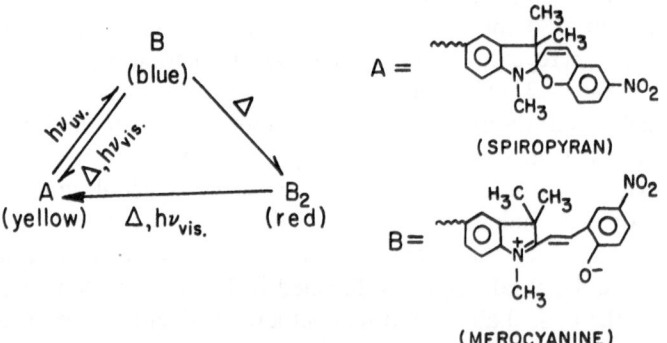

Figure 4.21 Scheme for the conversion of the photochromic side groups in the copolymer depicted in Figure 4.14 (Cabrera *et al.*, 1987).

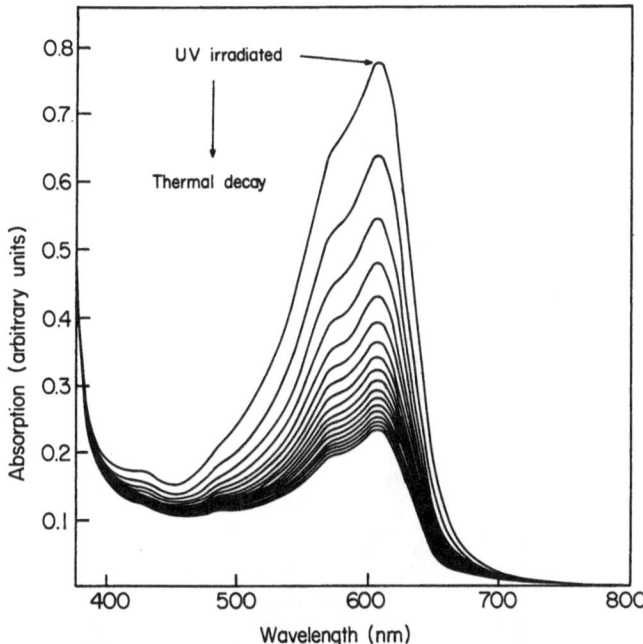

Figure 4.22 Change of absorption spectra of 12% copolymer P(l) (its structure is shown in Figure 4.11) during thermal color decay. Spectra were recorded at 20-min intervals after UV irradiation (Yitzchaik *et al.*, 1990a).

4.3.4.2. Thermal decay of photoinduced color. The main features of the spontaneous decoloration processes occurring in photochromic liquid crystal polymers are very similar to those in glassy polymers [see Krongauz (1990) and the references therein]. The color decay does not obey first-order kinetics and can be described only by at least biexponential equations (Cabrera *et al.*, 1987; Yitzchaik *et al.*, 1990a). For the merocyanine–spiropyran thermal conversion a hypsochromic shift in the course of the decay and deviation from first-order kinetics was explained by formation of dimers and higher merocyanine aggregates. The mesomorphic structure of the polymer glass does not significantly retard the color decay, the effect being much less marked than an increase of the polymer polarity. The decay rate decrease with increasing photochrome concentration was observed in the polymers containing spiroxazine groups (Figure 4.24) (Yitzchaik *et al.*, 1990a). The slowest decay was observed in the homopolymer containing only the spiroxazine side groups. The effect was attributed to steric hindrance of thermal ring closure by surrounding bulky photochromic groups, because the spiroxazines do not form aggregates. In more flexible polysiloxane polymers containing spiroxazine this effect was not observed (at least) at room temperature, which is well above T_g.

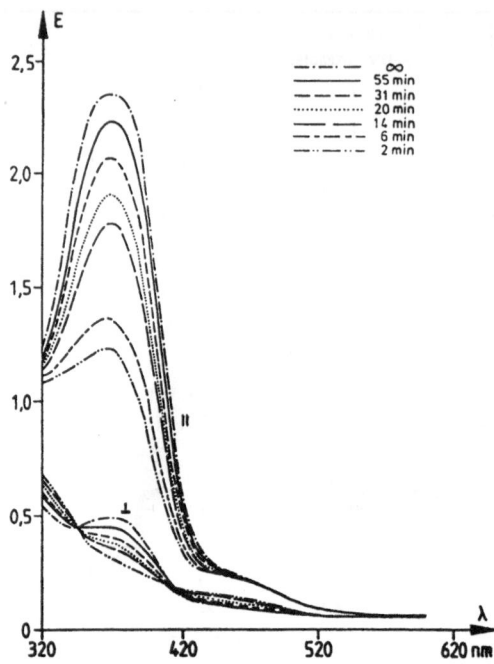

Figure 4.23 UV absorption spectrum for the azobenzene mesogenic units (the copolymer depicted in Figure 4.17) as obtained for a monodomain parallel and perpendicular to the director for various time intervals after irradiation. ∞ corresponds to the spectrum observed prior to the irradiation [1% (mol/mol) azobenzene comonomer units in acryl 0/100] (Wendorff and Eich, 1989).

4.3.4.3   Film alignment in an electric field and order parameter

The alignment of the photochromic liquid crystal copolymers in an electric field is quite similar to that of the corresponding non-photochromic liquid crystal homopolymers (see, for example, Ringsdorf and Zentel, 1982). It is convenient to align a polymer homogeneously between parallel electrodes deposited on a glass slide in order to measure the order parameter of the dye units. An electrostatic field of about 1 kV/mm is usually sufficient to align the liquid crystal polymer with the appropriate dielectric anisotropy homogeneously at a temperature above T_g. Cooling the aligned films below T_g in the electric field and subsequent UV irradiation produces dichroic absorption in the visible, which allows estimation of the orientational order of the dye from eqn (3).

The orientation of photochromic as well as permanent dyes depends on the shape of the dyes and their means of fixation. For example, the anthraquinone dye in the copolymer shown in Figure 4.13 gives the order parameter of 0.7 (Figure 4.25). The merocyanine formed on irradiation of

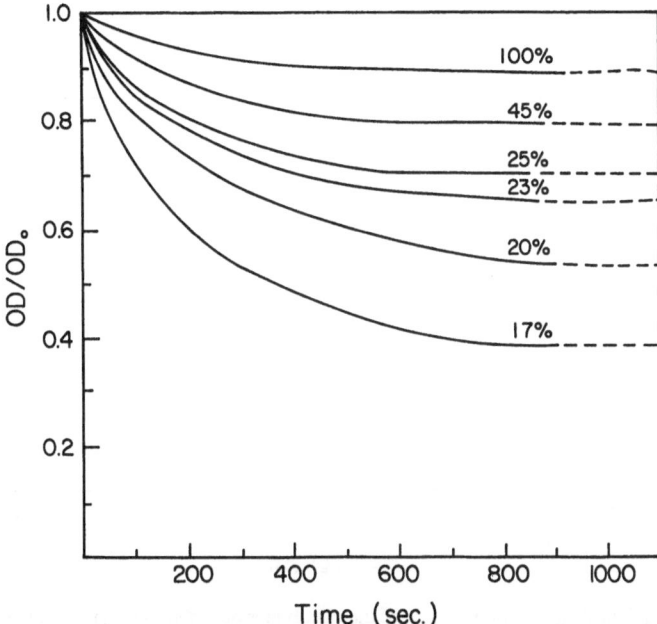

Figure 4.24 Thermal color decay of copolymer P(IV) (see Figure 4.11) with different spiro-oxazine content (indicated in the figure). Two upper curves relate to glassy amorphous films (Yitzchaik *et al.*, 1990a)

copolymers with spiropyran groups gives $S \approx 0.1$–0.2, while 4-dimethylamino,4′-nitrostilbene (DANS) dissolved in the same copolymers gives a substantially larger $S = 0.5$, probably because of the better compatability with mesogenic domains. The low-order parameter for the merocyanine groups formed from spiropyran was explained by the assumption that the photochromic groups are accommodated outside mesogenic domains. This is also confirmed by the rather weak influence of the mesophase on decoloration kinetics and the absence of impact of photochromic conversion on the film alignment.

The hypothesis of the existence of two sites in photochromic side-chain liquid crystal polymers (mesogenic domains and amorphous sites) implies that polymer backbones and photochromic side chains are located in the amorphous site, which is expanded with an increase of the photochrome content until mesophase eventually vanishes. However, a certain 'mixing' of mesogenic domains and spiropyran groups occurs which is manifested in a non-zero order parameter and certain retardation of thermal color decay in some copolymers. The rheo-optical effect discussed in the next section also indicates the existence of indirect interactions which apparently involve the main chain.

4.3.4.4. The rheo-optical effect. A remarkable feature of the isotropic

$$S_D = \frac{A\| - A\perp}{A\| + 2A\perp}$$

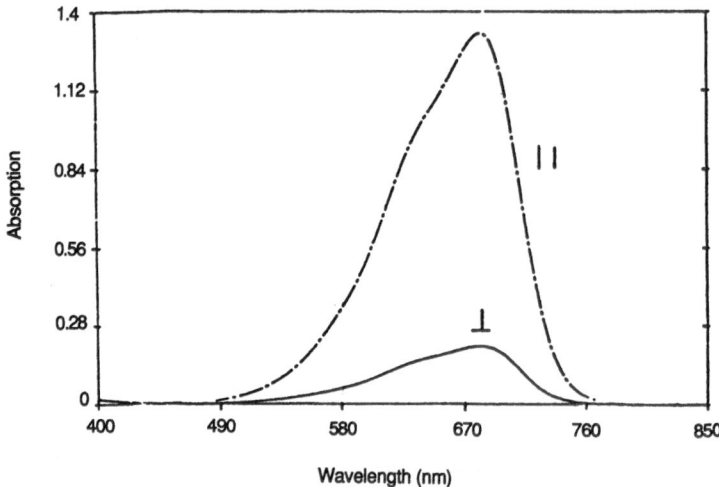

Figure 4.25 Dichroic absorption spectra of the copolymer depicted in Figure 4.13 [8% (w/w)] (Ringsdorf *et al.*, 1985).

films formed above the clearing point by the copolymers with low concentration of spiropyrans is their very strong transient translucence between crossed polarizers when they are squeezed between two glass slides or even lightly touched with the tip of a spatula (Cabrera and Krongauz, 1987a, b). The liquid crystal homopolymer which does not contain spiropyran, does not exhibit this effect.

Usually such instant birefringence during mechanical disturbance is considered as an indication of homeotropic (orthogonal to a solid surface) orientation of mesogenic molecules (Kelker and Hatz, 1980). Therefore one could conclude that instead of a nematic–isotropic transition at the clearing point a transformation of orientation of mesogenic groups of macromolecules from parallel to perpendicular to the surface is observed.

To check this hypothesis, the glass surface was treated with cremophor and nylon which promote the planar orientation of liquid crystals, and with 1-dodecanol for homeotropic alignment (De Jeu, 1980; Patel *et al.*, 1984). No effect of the surface on the 'sparkling phenomenon' was found. A decisive experiment was performed with droplets of the copolymers in an isotropic phase, having a relatively small contact area with a surface. With these droplets again even a very gentle touch gave remarkable sparkling, although in this case the effect of the surface must be insignificant. This suggests that the transient brightening is caused by at least partial restoration of liquid crystalline order induced by mechanical disturbance.

A parallel-disc-type rheometer with transparent glass discs was built for measurements of these rheo-optical properties of a film as a function of temperature.

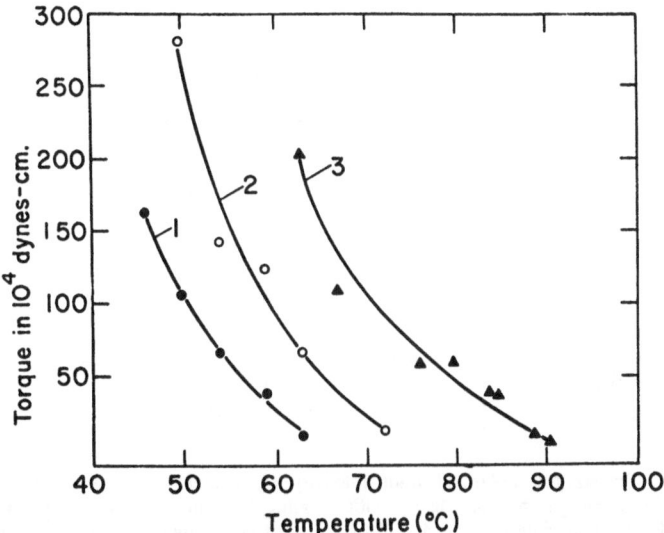

Figure 4.26 The temperature dependence of the viscosity represented as torque required to rotate the rheometer in the liquid crystal polyacrylates depicted in Figure 4.16 ($n = 5$). 1, liquid crystal polyacrylate homopolymer ($x = 0$); 2, 10% copolymer ($x = 10$, $y = 90$); 3, 22% copolymer ($x = 22$, $y = 78$).

The torque required to rotate the upper disc of the rheometer was measured as a function of the polymer melt temperature (Figure 4.26). The torques of both the liquid crystalline homopolymer and copolymers decrease sharply above the glass transition points. Since the viscosity is a monotonically increasing function of the torque (Bird *et al.*, 1977), one can conclude that the viscosity of the copolymers is much higher than that of the homopolymer at the same temperature. The viscosities of spiropyran copolymers with long spacers [$-(CH_2)-_5$ and $-(CH_2)-_{11}$] are very similar, while the copolymers with the spacer $-(CH_2)-_2$ have a much higher viscosity.

A drastic increase of viscosity on the introduction of a relatively small portion of the thermochromic groups indicates a very substantial effect of these groups on the interaction between macromolecules. One may assume that the interaction of the merocyanine groups, formed on heating, gives rise to aggregation of macromolecules into a network with high viscosity. Apparently the copolymers in which the spiropyran groups are separated from the backbone by short spacers give a much denser network than the copolymers with the 'long spacer', resulting in the higher viscosity of the melt.

The polarized light intensity–temperature relationships for the liquid crystalline homopolymer and the spiropyran copolymers are shown in

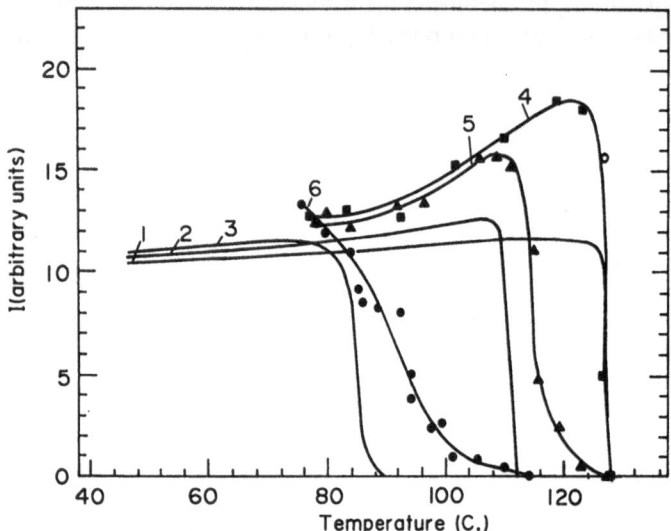

Figure 4.27 Transmitted polarized light intensity (I) against temperature measured in rheometer with a transparent glass disc: 'Static' regime 1, homopolymer; 2, 10% (mol/mol) copolymer depicted in Figure 4.16; 3, 22% (mol/mol) copolymer; 4–6 as 1–3, but in the 'dynamic' regime.

Figure 4.27. The clearing points are indicated by a sharp fall of 'static' birefringence. In the dynamic regime the transmitted light intensity reaches a maximum near the clearing points. This can be explained by a decrease of the number of defects with viscosity decrease and domain growth induced by shear (Asada *et al.*, 1980). At low temperatures the viscosity becomes so high that it is impossible to perform the shearing measurements. The viscosity also increases with the spiropyran content, for example 46% polymer forms a red, tar-like material on melting, which gives neither a mesomorphic phase nor dynamic birefringence.

The dynamic birefringence of the homopolymer disappears at the clearing point almost as sharply as does the static one, while that of the copolymers extends much beyond the clearing points, though it decreases gradually beyond these points. The range between the clearing point and the temperature where the dynamic birefringence is no longer observed increases with the spiropyran content of the copolymers.

In other words, even though the spiropyran–merocyanine groups disturb the mesophase thermal stability, they promote the dynamic restoration of an ephemeral order in the isotropic phase.

Films of the copolymers cast onto a glass slide with interdigital electrodes can be aligned in an electrostatic field (> 0.5 kV/mm) below the clearing point. Above this temperature the alignment disappears. This, together with the DSC measurements, confirms that the disappearance of static birefringence relates to the mesophase–isotropic phase transition.

Apparently physical cross-linking of the copolymers due to aggregation of the merocyanine groups induces the rheo-optical effect. In order to stimulate the merocyanine aggregation at room temperature the copolymer film was swelled in chloroform. After swelling and drying the static birefringence disappeared but a very distinct dynamic birefringence was observed at room temperature. The rheo-optical effect was also observed at room temperature in polysiloxane copolymers bearing spiropyran side groups.

To explain how the network formation leads to the rheo-optical effect one has to consider the effect of the liquid crystal–isotropic phase transition on the main chain conformation. According to Kuzma (1983), in the liquid crystal phase the parallel arrangement of mesogenic groups affects the conformation of the macromolecular backbone. Above the clearing point this effect of the mesogenic side groups becomes insignificant and the macromolecules adopt a coil conformation, which provides maximum entropy, and which is in general different from the conformation in the mesophase. Formation of the transient liquid crystalline order by shearing or other mechanical perturbation results not only in restoration of the parallel arrangement of the mesogenic groups but in a certain reconstruction of the main-chain coils, causing a decrease in entropy.

Presumably, formation of the network favors the preservation by macromolecules of the conformation acquired in the mesophase even above the clearing point because of a more rigid structure. This makes the dynamic ordering easier.

Earlier Finkelman et al. (1984) observed the occurrence of birefringence above the clearing point on stretching side-chain liquid crystal polysiloxane elastomers. This observation is consistent with our conclusion that dynamic birefringence stems from macromolecular arrangement.

4.3.4.5 Blends of photochromic liquid crystal polymers and low molar mass liquid crystals (Yitchaik *et al.*, 1990c). The phase diagram of mixtures of the photochromic liquid crystal copolymer depicted in Figure 4.16 ($n = 5$, $x{:}y = 1.4$) and a low molecular weight nematic liquid crystal, 4-(cyanophenyl)-heptylbenzoate, is shown in Figure 4.28. The diagram is very similar to that reported by Ringsdorf *et al.* (1982) for the liquid homopolymer (which does not contain photochromic units) mixed with the same liquid crystal. The copolymer is plasticized (T_g goes down) by the liquid crystal, while crystallization of the latter is markedly depressed at high copolymer content. Full miscibility of the components was observed in the nematic phase, while it is restricted to ~ 25% of liquid crystal in the glassy state. One could expect that the miscibility of the copolymer with a QLC should be better than with a conventional liquid crystal because QLC is composed of two units structurally similar to both comonomers of the copolymer. Indeed, the phase diagram of the copolymer–QLC blends (Figure 4.29) shows full miscibility of these two mesomorphic materials below the clearing temperature. Metastability of the nematic blends in the region D is determined mainly by the inherent metastability of the QLC mesophase itself. Fast preparation and cooling of the blend to below T_g prevented phase separation and gave a uniform glassy film (region F). Conversely, annealing of the mesophase above T_g accelerated phase segregation apparently as a result of merocyanine aggregation. In fact, the boundary between regions C and D and between E and F should be considered as only approximate, because one cannot exclude that on longer annealing further aggregation and crystallization of QLC might occur.

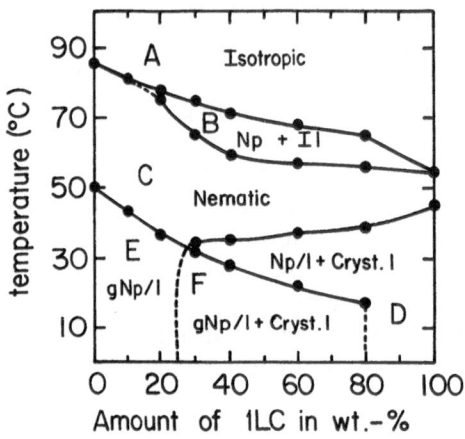

Figure 4.28 Phase diagram of blends of the copolymer depicted in Figure 4.16 with 4-(cyanophenyl)-heptylbenzoate (gNp, glassy nematic; Np, nematic; I, isotropic).

Unlike the low-molar-mass liquid crystal which plasticized the polymer, blending of the copolymer with QLC increased T_g. Apparently the free volume of the blends is lower, and their structure is more rigid than that of the copolymer because QLCs are more compatible both with mesogenic and photochromic groups.

The photochromic properties of the blends are not very different from those of the copolymer alone. UV irradiation of a glassy blend film led to the appearance of a visible absorption band with $\lambda_{max} \approx 590$–595 μm, which is slightly red-shifted with respect to the $\lambda_{max} \approx 580$–585 of the UV-irradiated copolymer. Annealing the blends above T_g without irradiation resulted in a broad absorption plateau in the range of 480–570 nm, which is similar to the spectrum obtained for the copolymer itself via thermochromic reaction. The spectrum was ascribed to merocyanine aggregates. The thermal decay of the photoinduced color does not obey first-order kinetics and proceeds faster with increased QLC content.

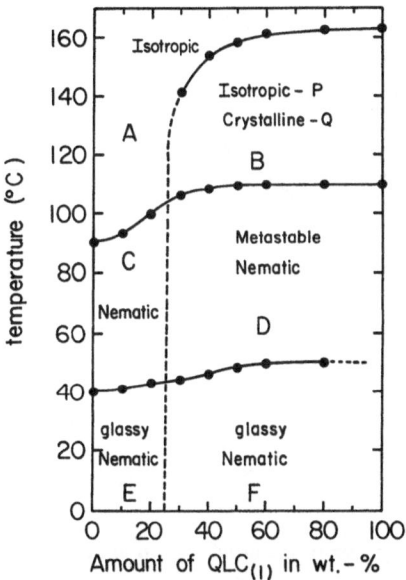

Figure 4.29 Phase diagram of blends of the copolymer (see Figure 4.16) with QLC (N 9 in Table 4.2).

The order parameter of merocyanine formed on irradiation of the blend films aligned in an electrostatic field as a function of blend composition is presented in Figure 4.30. The merocyanine groups belong mainly to the copolymer because the QLC component is not photochromic and contributes substantially less to the color formation. The fact that blending with QLC substantially improved the merocyanine alignment is consistent with the assumption about compatibility of QLCs with both mesogenic and photochromic groups. Some improvement of the alignment was also observed for the blends of copolymer and low molar mass liquid crystal.

4.4 Non-linear optical properties

4.4.1 Basic concepts

Non-linear optical properties of organic compounds and polymers have become a topic of active study during the last 10 years. Several publications on this topic deserve to be mentioned (Williams, 1983, 1984; Prasad and Ulrich, 1988; Hann and Bloor, 1989; Lyons, 1989; Shen, 1984).

Polarization (P) induced in an atom or molecule by an external field (E) can be written as:

$$P = \alpha E + \beta E^2 + \gamma E^3 + \ldots \tag{4}$$

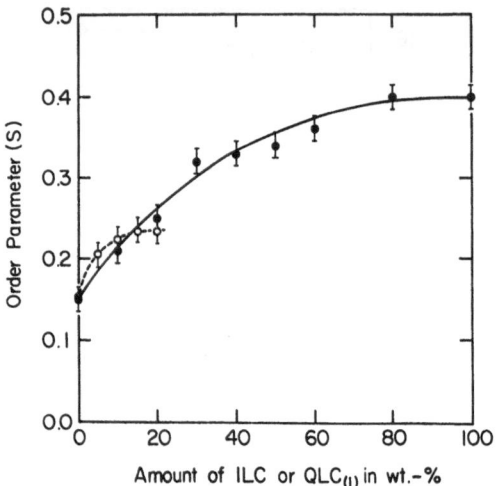

Figure 4.30 Order parameter of the blends shown in Figures 4.28 (o) and 4.29 (•).

where the tensor quantities α, β and γ are referred to as the polarizability (α), hyperpolarizability (β), etc. A similar equation for bulk media can be expressed as:

$$P = \chi^{(1)} E + \chi^{(2)} E_2 + \chi^{(3)} E_3 + \ldots \qquad (5)$$

where $\chi^{(1)}$ is the linear susceptibility, $\chi^{(2)}$ is the non-linear or second-order susceptibility, etc. From the perspective of symmetry it is apparent that the even-order tensors β and $\chi^{(2)}$ are zero in centrosymmetric media, while the odd-order tensors α, $\chi^{(1)}$ and γ, $\chi^{(3)}$ do not have the symmetry restrictions. At the molecular level this means that only molecules with non-centrosymmetric structure can exhibit non-zero hyperpolarizability (β). It does not necessarily mean that at the macroscopic level the material composed of non-centrosymmetric molecules has non-zero $\chi^{(2)}$. Molecules with an asymmetric charge distribution and therefore a non-zero β may exist in a centrosymmetric crystal or in an orientationally averaged molecular environment such as a liquid or amorphous polymer. Therefore, they may exhibit an extremely small $\chi^{(2)}$.

Two approaches have been developed to produce non-centrosymmetric orientation in non-crystalline media: (1) poling of organic and in particular polymeric films in an electrostatic field and (2) formation of molecular monolayers and multilayers on interfaces (Langmuir–Blodgett or self-assembled films).

Poling of polymeric films containing a dipolar chromophore of high hyperpolarizability (β) in an external DC electric field is a common method for achieving a non-centrosymmetric structure and large second-order non-linear coefficients (Williams, 1984; Singer *et al.*, 1986; Boyd, 1989; Zyss, 1985; Lylel *et al.*, 1987). This can be accomplished by heating the polymer film above glass transition temperatures followed by cooling below T_g in an electrostatic field. The polar alignment of the high β-molecules might be frozen in the rigid glassy polymer (Figure 4.31).

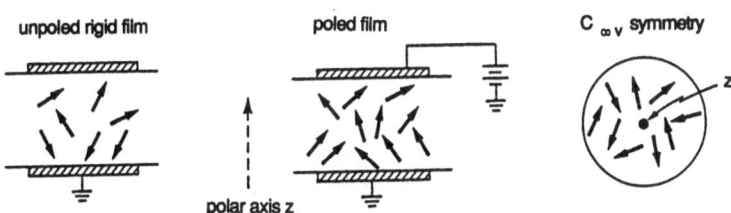

Figure 4.31 Schematic of imparting $C_{\infty v}$ symmetry (polar axis, isotropic perpendicular to the polar axis) to an isotropic medium containing permanent dipoles with an external electric field E (Williams, 1984).

The second-order non-linearity leads to a number of important optical effects such as second harmonic generation, Pockel's effect, optical rectification, etc. Here only second harmonic generation (SHG) will be considered, which can be understood with a simple explanation (Shen, 1984). In terms of components, the second-order polarization $P^{(2)}$ can be expressed as follows:

$$E(t) - E_0 e^{i\omega t} \tag{6}$$

The interaction with two identical optical fields results in:

$$P^{(2)} \sim e^{i2wt} \tag{7}$$

i.e. an optical field of twice this frequency. This is the phenomenon of second harmonic generation (SHG) whereby some of the incident laser light is converted to light of twice the frequency.

If a DC electric field (time-independent) is used in conjunction with an optical field, the non-linear polarization, $P^{(2)}$, retains the frequency dependence of the input laser. However, the $\chi^{(2)}$ term may induce a phase change (refractive index change) or a polarization direction change of $P^{(2)}$ relative to the input laser. These are examples of the electro-optic (EO) effect.

4.4.2 Theory of SHG from a mesophase containing high-β dye (Hsiung et al., 1987).

A non-linear polarization P, of frequency $2\tilde{\omega}$, can be induced in a medium irradiated by a laser field E, of frequency ω, as described by the relation:

$$P(2\tilde{\omega}) = \chi^{(2)}(2\tilde{\omega} = \tilde{\omega} + \tilde{\omega}) : E(\tilde{\omega})E(\tilde{\omega}) \tag{8}$$

P then acts as a source that radiates coherent waves at 2ω. For a medium exhibiting complete rotational symmetry about a polar axis, $\hat{x}$ (i.e. $C_{\infty v}$ symmetry), the non-zero elements of $\chi^{(2)}$ are $\chi^{(2)}_{xxx}$, $\chi^{(2)}_{xyy} = \chi^{(2)}_{xzz}$, and $\chi^{(2)}_{yyx} = \chi^{(2)}_{yxy} = \chi^{(2)}_{zzx} = \chi^{(2)}_{zxz}$. In an organic substance where the identities of individual molecules are preserved, the macroscopic $\chi^{(2)}$ can be related to the molecular hyperpolarizability $\beta(2\tilde{\omega} = \tilde{\omega} + \tilde{\omega})$ through a statistical averaging over molecular orientations:

$$\chi^{(2)} = N \langle \beta: L(2\tilde{\omega})L(\tilde{\omega})L(\tilde{\omega}) \rangle, \tag{9}$$

where N is the number density and the Ls are the local field correction factors (Bloembergen, 1965) which account for the molecular interactions in the substance. In the later analysis, the well-known Lorenz–Lorentz form will be used, $L(\tilde{\omega}) = [\varepsilon(\tilde{\omega}) + 2]/3$, ε being the linear dielectric tensor of the medium.

Since high-β dyes are known to have very large molecular hyperpolarizabilities ($\beta \approx 10^{-28}$ to 10^{-27} esu), (Williams, 1984) it can be assumed that SHG in the dye-containing mesophase is dominated by the dye molecules. For simplicity, it is further assumed that both β and the permanent dipole moment p of a dye are dominated by a single component along the long molecular axis ($\hat{\xi}$), that is by $\beta_{\xi\xi\xi}$ ($\equiv \beta$) and $p_\xi (\equiv p)$ respectively. The validity of these assumptions will be discussed later. Then, from eqn. (9):

$$\chi^{(2)}_{xxx} = N\beta L_x(2\tilde{\omega})[L_x(\tilde{\omega})]^2 \langle \cos^3 \theta \rangle,$$

$$\chi^{(2)}_{xyy} = N\beta L_x(2\tilde{\omega})[L_y(\tilde{\omega})]^2 [\langle \cos \theta \rangle, - \langle \cos^3 \theta \rangle]/2, \tag{10}$$

$$\chi^{(2)}_{yyx} = N\beta L_y(2\tilde{\omega})L_y(\tilde{\omega})L_x(\tilde{\omega})[\langle \cos \theta \rangle - \langle \cos^3 \theta \rangle]/2$$

where θ is the angle between ξ and $\hat{x}$, and N now refers to the number density of the dye. For the averaging over molecular orientations, it is assumed that the 'mean field' potential experienced by an individual merocyanine molecule in a QLC in the presence of an external DC field, $E_0(= E\hat{x})$ is:

$$V(\theta) = V_N(\theta) - u \cos \theta - pL_0E_0$$

$$V_N(\theta) = -u_2 P_2(\cos \theta) - u_4 P_4(\cos \theta) - \ldots \tag{11}$$

where $L_0 \equiv L(\tilde{\omega} = 0)$, P_n is the nth-order Legendre polynomial, and V_N is the conventional nematic mean field potential (Priestly et al., 1975). The coefficients u and u_n are determined by molecular ordering and interaction strength. The term $u \cos \theta$ is included here to account for the possible polar symmetry of a mesophase even in the absence of an external DC field. Both this term and the pL_0E_0 term will be considered as perturbations to the nematic potential in the present case. The orientation-averaged value of an arbitrary function $f(\theta)$ can then be obtained from:

$$\langle f(\theta)\rangle = \int_{-1}^{1} f(\theta) \, \exp[-V(\theta)/kT] d(\cos\,\theta)/$$

$$\int_{-1}^{1} \exp[-V(\theta)/kT] d(\cos\,\theta) \tag{12}$$

$$\approx \langle f(\theta)\rangle_0 + \langle f(\theta) \cos\,\theta\rangle_0 (u + pL_0E)/kT$$

where:

$$\langle g(\theta)\rangle_0 \equiv \int_{-1}^{1} g(\theta) \, \exp[-V_N(\theta)/kT] d(\cos\,\theta)/$$

$$\int_{-1}^{1} \exp[-V_N(\theta)/kT] d(\cos\,\theta)$$

Since $V_N(\theta)$ is an even function of $\cos\,\theta$:

$$\langle \cos^n\,\theta\rangle = \langle \cos^n\,\theta\rangle_0 \text{ for even } n$$

$$\langle \cos^{n+1}\,\theta\rangle_0 \,(u + pL_0E)/kT \text{ for odd } n \tag{13}$$

From eqns (10) and (13), notice that $\chi^{(2)}$ should depend linearly on E, while the ratio between any two $\chi_{ijk}^{(2)}$s should remain constant. On the other hand, the nematic-type order parameter:

$$S \equiv \langle P_2(\cos\,\theta)\rangle \equiv (3\langle\cos^2\,\theta\rangle - 1)/2 = \langle P_2(\cos\,\theta)\rangle_0 \tag{14}$$

should be independent of the E field.

4.4.3 Non-linear optical susceptibility measurements

The first measurements were conducted on side-chain liquid crystal polymer films doped with DANS and aligned in a DC field (Meredith et al., 1982). Later QLCs were investigated in a similar way (Hsiung et al., 1987). Recently more comprehensive studies of SHG by QLCs, photochromic liquid crystal polymers containing spiropyran and merocyanine groups, and by these polymers doped with different high-β species were performed (Yitzchaik et al., 1990 a, b, d). These studies revealed that along with the conventional optical non-linearity, which is described by the theory, a two-dimensional optical non-linearity appears, following electrical field poling along one direction only. Consequently, these films exhibit more non-zero components of the second-order susceptibility tensor ($\chi^{(2)}$) than were observed earlier. First a detailed description of conventional non-linear optical susceptibility measurements performed on QLC films is presented below as an example (Hsiung et al., 1987).

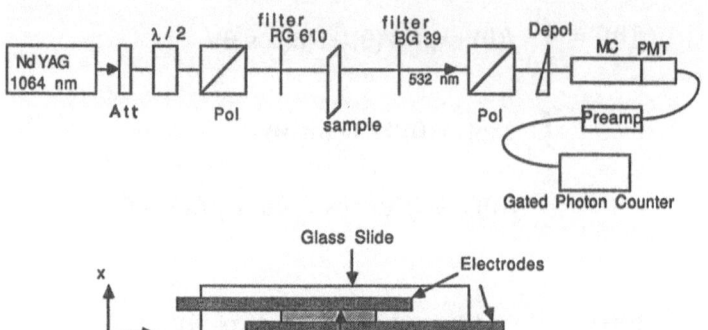

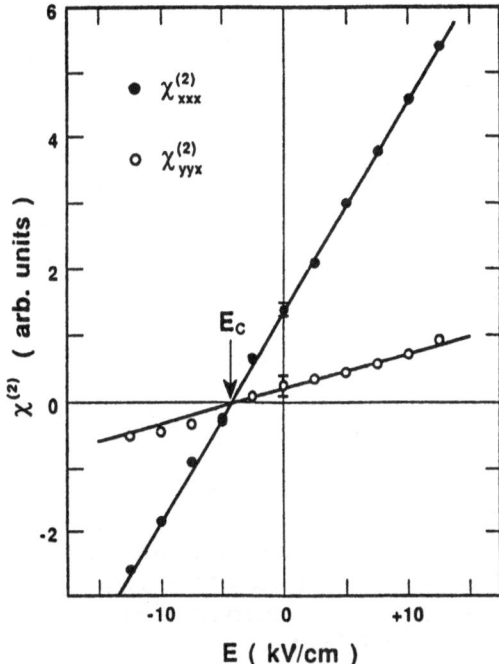

Figure 4.33 Two tensor components of the non-linear susceptibility $\chi^{(2)}$ as a function of the strength of an electrostatic field applied along the director (X) of a QLC film.

This effect is presumably due to relaxation of molecular orientations in QLC towards a metastable orientational distribution.

The non-vanishing $\chi^{(2)}$ in the glassy QLCs after removal of the external field indicates that they are non-centrosymmetric. The E field-dependence of $\chi^{(2)}_{ijk}$s, shown in Figure 4.33, agrees with eqns (10) and (13) if $u = -pL_0E_c$. The quantitative agreement with the theory was checked as follows. From the experimental values of $\chi^{(2)}_{yyx}/\chi^{(2)}_{xxx}$ and $\langle\cos^2\theta\rangle = (2S + 1)/3$, with $S \approx 0.4$, it can be deduced from eqns. (10) and (13) that $\langle\cos^4\theta\rangle_0 \approx 0.45$, i.e. $\langle P_4(\cos\theta)\rangle \approx 0.09$, a reasonable value for the nematic ordering (Durbin and Shen, 1984). Taking the number density N of merocyanines to be $\sim 2 \times 10^{19}/\text{cm}^3$ ($\sim 5\%$ of the total density) and their molecular dipole moment $p \sim 10$ D (Bergmann *et al.*, 1950), the optical dielectric constants $\varepsilon_x \sim 2.6$ and $\varepsilon_y \sim 2.3$ (neglecting optical dispersion), the DC dielectric constant $\varepsilon \sim 10$–20, and using the experimental values of $\chi^{(2)}_{xxx}$ and E_c, then from eqn (10) $\beta \sim 1$–2×10^{-28} esu for the QLC merocyanines. This value is of the same order of magnitude as those of other merocyanines (Williams, 1984).

The experimental value of $\chi^{(2)}_{xyy}/\chi^{(2)}_{yyx}$ is ≈ 0.6. From eqn. (10), this suggests that $L_x(2\tilde{\omega})L_y(\tilde{\omega})/L_y(2\tilde{\omega})L_x(\tilde{\omega}) \sim 0.6$. This could be the result of local field dispersion due to a resonance band near $2\tilde{\omega}$ in merocyanines. On the other hand, it may also be the consequence of the simplifying assumption that all but one component of β is negligible. Spiropyrans, because of their much larger concentration in QLCs, may also contribute to the SHG even though their β and p are much smaller than those of the merocyanines. Such a correction, however, can be easily included in the theory. The opposite polarities of $\chi^{(2)}$ $(E = 0)$ (or E_c) observed in different samples should correspond to the two metastable states of the polar ordering with their symmetry axes pointing in the $+$ and $-\hat{x}$ directions. The non-linear dependence of E in the case of 'negative $\chi^{(2)}$ $(E = 0)$' indicates that the two states are non-degenerate; the negative $\chi^{(2)}$ $(E = 0)$ state is less stable and more easily influenced by the applied field.

It is possible that in addition to molecular reorientation a pure electronic process described by:

$$P^{NL}(2\tilde{\omega}) = \chi^3(2\tilde{\omega} = 0 + \tilde{\omega} + \tilde{\omega}): E_0 E(\tilde{\omega}) E(\tilde{\omega}) \qquad (15)$$

may also contribute to the field-induced part of the SHG. However, for this contribution to be important, χ^3 in eqn. (15) must have a value $\sim 2\text{--}6 \times 10^{-12}$ esu, which seems to be too large for typical molecules. Therefore, the pure electronic contribution is not likely to be very significant.

Polar ordering has never been observed in conventional liquid crystals in the nematic state, that is molecules tend to have equal probabilities of pointing in opposite directions along the director. Local antiferroelectric pairing of molecules is also known to exist in many liquid crystals consisting of strong polar molecules. This is because of a tendency for nearby dipoles to prefer to be antiparallel to each other in order to reduce their dipole–dipole interactions (Chandrasekhar, 1977). Apparently in the QLCs the merocyanine molecules are also partially aggregated. However, merocyanine molecules are at least partially separated by a few spiropyrans, which could screen the dipole–dipole interactions among merocyanines and allow a preferred polar alignment of these molecules; in other words, the spiropyrans are responsible for the interaction potential $-u \cos \theta$ given in eqn. (11).

4.4.4 Two-dimensional asymmetry and optical non-linearity upon poling by an electrostatic field (Yitzchaik et al., 1990c–f)

This effect was observed when a merocyanine-containing mesomorphic film (QLC or an acrylic liquid crystal copolymer with a spiropyran) was cast on a glass slide onto which thin-film metal electrodes had been deposited. The sample was rotated around vertical axis x (Figure 4.32) under applied electrostatic field at room temperature. In this case, in addition to the signal where both the input fundamental and generated harmonic are vertically polarized (I_{v-v} signal), very strong I_{h-h} (i.e. horizontally polarized input and output) and I_{v-h} (vertical input, horizontal output) signals appeared when the sample was for example at 45° to the beam. The I_{h-v} signal however, remained very weak.

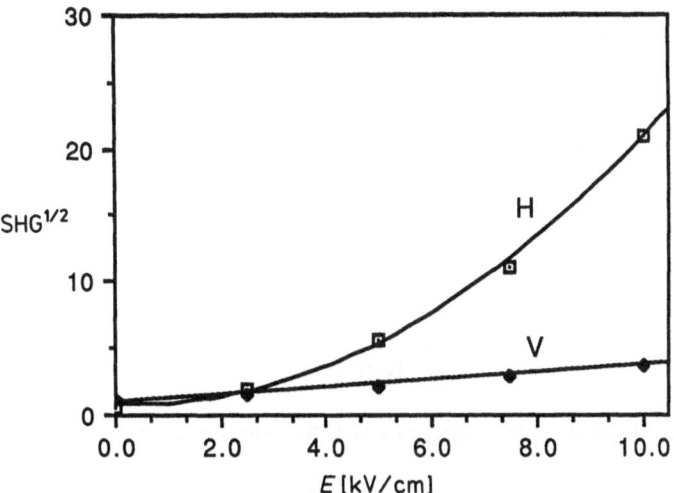

Figure 4.34 Dependence of $I_{v-v}^{1/2}$ (V) and $I_{h-h}^{1/2}$ (H) signals of a QLC on the applied electric field. The $L_{v-v}^{1/2}$ data are fitted by a linear dependence on the field; $I_{h-h}^{1/2}$ is fitted by a quadratic dependence.

The dependence of the I_{h-h} and I_{v-v} signals on the electric field was different (Figure 4.34); I_{v-v}, the only significant signal in the absence of the DC field ($I_{v-v}^{1/2} \sim X_{xxx}^{(2)}$), increased relatively slowly and linearly with the electric field (see also Figure 4.33) as this was expected for interaction between isolated dipole moments and a DC field. The I_{h-h} signal showed a much more dramatic increase with the field. In the range studied $I_{h-h}^{1/2}$ ($\sim \chi_{zzz}$) was proportional to the square of the field.

The I_{h-h} slowly disappeared after the DC field was removed, decaying in a first-order manner over about an hour in a QLC sample (Figure 4.35). Thus the DC field clearly induced a transient asymmetry in either the y or z direction. Since the strong I_{h-h} and I_{v-h} did not appear when the sample was normal to the beam (i.e. the horizontal polarization direction parallel to y), this verified that no asymmetry was induced in the y direction. These strong signals could arise only from an induced asymmetry in the z direction (normal to the plane of the sample), meaning that the I_{v-h} signal arises from the $\chi_{zxx}^{(2)}$ component and the I_{h-h} signal must arise from among the $\chi_{zzz}^{(2)}$, $\chi_{zyy}^{(2)}$ and $\chi_{yzy}^{(2)}$ components. By carrying out analogous measurements at 45° incidence but with the y axis vertical, it was concluded that the dominant contributions to the SHG signal come via $\chi_{zzz}^{(2)}$. Elucidation of the process responsible for the I_{h-h} signal when the DC electric field is applied may be

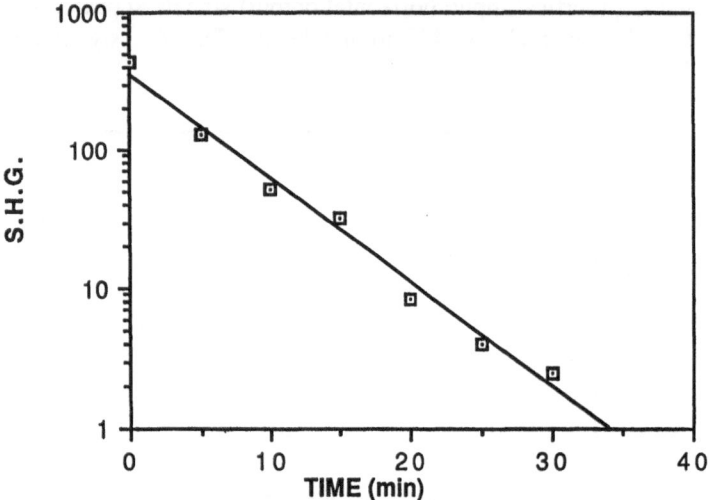

Figure 4.35 Decay of I_{h-h} SHG signal of a QLC following the removal of the electric field.

assisted by experiments with photochromic liquid crystal polymers containing spiropyran–merocyanine comonomers and with blends of these polymers with QLCs. Doping with stable dipolar dyes such as DANS was also examined. Some results on SHG from these systems are presented in Table 4.4 (the photochromic liquid crystal polymer depicted in Figure 4.16 was used, which is denoted by PLCP in Table 4.4).

In the absence of an electric field (the field was removed after the poling procedure was completed) the freshly prepared samples showed SHG where a dominant contribution was $\chi^{(2)}_{xxx}$ (signal I_{v-v}). The intensity of I_{v-v} increased in the sequence QLCs <PLCP<PLCP–QLC blends <PLCP–QLC–DANS blends. No substantial I_{h-h} signal was observed. However, on application of the field as described above, I_{h-h} signals appeared in which $I_{v-v} < I_{v-h} < I_{h-h}$. Comparison of the freshly prepared and 'deactivated' several weeks' old samples shows that, while the I_{v-v} is negligible in the old sample even when the strong field is applied, the I_{h-h} signals are strong and comparable both in the fresh and old samples. The I_{h-h} signal was observed even in the samples prepared without UV irradiation. According to the results described above (see sections 4.3.3 and 4.3.4.2) the I_{h-h} signal can come only from merocyanine dimers or higher aggregates. The non-aggregated merocyanine units (i.e. isolated dipolar groups) have either decayed (the 'old' samples) or were not formed at all (the polymer sample prepared without irradiation). The merocyanine groups in such samples are preserved only in the aggregated form.

The aggregates are of the H type and are expected to have antiparallel dipole arrangements (Figure 4.36a). Application of the DC field may bring

Table 4.4 Relative SHG signals measured in transmission[a].

	No field[b]	In DC electric field 1× 10⁴ V/cm[c]		
	I_{v-v}	I_{h-h}	I_{v-v}	I_{v-h}
PLCP–QLC(1) blend 4:1 — fresh[d]	3–10	$6 \pm 2 \times 10^2$	20–3	1×10^2
PLCP–QLC(2) blend 4:1 — old	< 0.1	1×10^2	5	6
QLC(1)–QLC(2) blend 1:3 — fresh	0.4–1.0	$6 \pm 3 \times 10^2$	20 ± 10	$1–2 \times 10^2$
QLC(1)–QLC(2) blend 1:3 — old	< 0.1	4×10^2	5	40
QLC(1)	0.3	$1–2 \times 10^2$	50	40
PLCP	1–2	40–60	5–10	
PLCP—not aligned	< 0.1	$2 \pm 1 \times 10^2$	3–5	15
PLCP—aligned without irradiation	< 0.1	30–40	2	
PLCP on PMMA	0.3	< 0.1	0.3	

[a] The plane of the sample was at 45° to propagation of the input laser beam. The x direction of the sample, i.e. the direction of the alignment by the DC electric field, was always vertical. I_{v-v} = signal for vertically polarized input, vertically polarized SHG output; I_{v-h} = vertically polarized input horizontally polarized SHG output, etc.
[b] In all samples, the other signal components, *viz.* I_{h-h}, I_{v-h} and I_{h-v} were not observed.
[c] In all samples no I_{h-v} signal was observed.
[d] QLC(1) and QLC(2) are compounds 9 and 10 in Table 4.2.

about some reorientation of individual dipoles within a stack, resulting in non-centrosymmetric structure (Figure 4.36b). Such a mechanism of electric field-induced SHG was demonstrated for another type of merocyanine aggregates, the so-called merocyanine quasicrystals (Krongauz *et al.*, 1978; Krongauz, 1979).

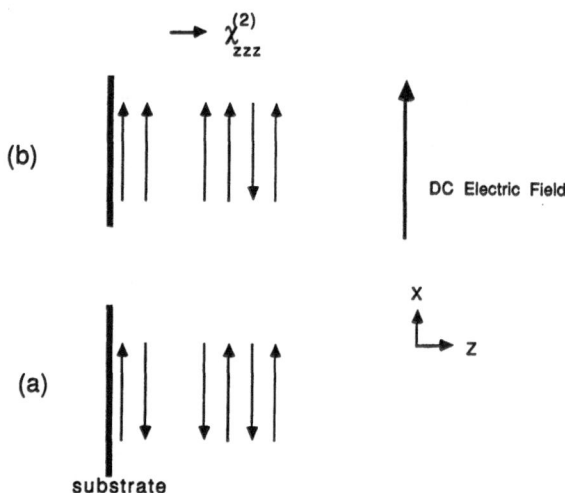

Figure 4.36 Proposed structure of merocyanine aggregates in the absence (a) and presence (b) of an electric field. The arrows represent the dipole moments of individual molecules.

However, aggregates such as depicted in Figure 4.36(b) may be responsible for the SHG signal arising only via $\chi^{(2)}_{xxx}$. In order to generate an SH signal polarized along the z axis, these aggregates must have asymmetry in the z direction. Presumably this is induced by some interactions between the glass substrate and the mesophase. A corollary of this, therefore, is that in the absence of such interactions the I_{h-h} signal should be absent. Some support for this model comes from the fact that no I_{h-h} signal was observed from the polymer film cast on a poly(methyl methacrylate) substrate. This was explained by the different substrate–mesophase interactions in the two cases (the copolymer on glass and the copolymer on PMMA), where only the former provides the interaction necessary to achieve an I_{h-h} signal.

Another possible explanation of the effect may be connected with the positioning of the thin metal electrodes in the interface between glass and the film. This may create a non-uniform electric field or non-uniform charge distribution due to the charge injection from the electrodes to the film (similar to the electret effect).

In conclusion, the non-zero $\chi^{(2)}$ coefficients could be exploited in electro-optical devices for phase-matched SHG and electro-optic modulation for light of different polarizations.

4.5 Optical information recording and storage

Liquid crystal side-chain polymers can be used for optical information thermorecording and storage (Shibaev *et al.*, in Blumstein, 1985; Coles and Simon, 1985; McArdle *et al.*, 1986). The principle of this recording is illustrated in Figure 4.37. A film of a nematic liquid crystal polymer is aligned at temperatures above T_g in an electric field homeotropically between transparent electrodes (Figure 4.37). The alignment is frozen in and stored at temperatures below T_g without the field. The film looks non-scattering and non-birefringent in the direction of alignment. The recording is carried out by local heating with a focused laser beam above the clearing temperature, T_c. This leads to local disorientation of the mesophase and appearance of a polydomain texture after cooling which is birefringent and scatters light. To erase the recorded information, the film should be heated above T_g and aligned again in the field. Reversibility of the recording is the major feature of the mesomorphic media.

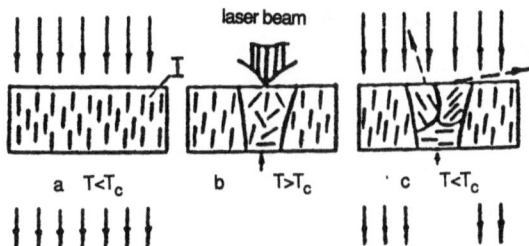

Figure 4.37 Scheme of thermorecording using the film of the homeotropically oriented nematic LC polymer. Only the mesogenic groups are shown (Shibaev *et al.*, 1985).

An optically clear nematic glass can also be obtained without an electric field by homogeneous planar alignment through annealing of a mesomorphic polymer on a surface which promotes such alignment (glass, polyimide, etc.). The sensitivity of the recording material can be substantially increased by incorporation of dye molecules (dissolved or covalently attached) that enhance the light absorption (Schmidt, 1989).

Recently, a few successful attempts to employ a photochemical reaction in a liquid crystal polymer for optical recording have been reported (Eich *et al.*, 1987a, b; Pinsl *et al.*, 1987; Ortler *et al.*, 1989; Wendorff and Eich, 1989). Cyclic polysiloxanes of the type represented in Figure 4.38, with side chains containing cholesteryl and mesogenic groups (R^1 and R^2) were doped with benzophenone or carbon black. The materials form a cholesteric phase, reflecting light whose wavelength in visible or near-IR depends on the relative content of R^1 and R^2. Irradiation of the polymer films with a visible laser beam produced irreversible destruction of the reflection ability. The reflection change which reached 80% was apparently produced by chemical reaction of a dopant with the polymer (in the case of benzophenone this presumably was an interaction of the ketyl radical with the matrix). The materials allowed non-destructive reading of the optical contrast with the low-intensity beam because the chemical reaction has a power threshold.

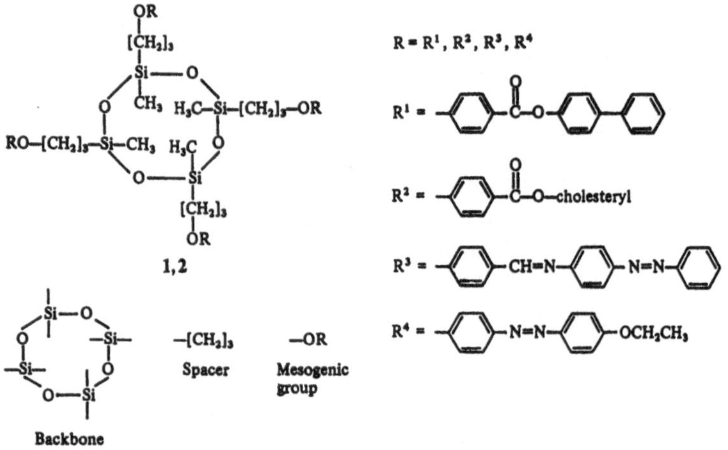

1: Mole ratio R^1 : R^2 : R^3 = 2 : 1,6 : 0,4; T_g = 50°C; T_c = 192°C; λ_{ref} = 790 nm

2: Mole ratio R^1 : R^2 : R^4 = 2 : 1,8 : 0,2; T_g = 49°C; T_c = 206°C; λ_{ref} = 695 nm

Figure 4.38 Photochromic cholesteric cyclic polysiloxane used for optical holographic M-recording (Ortler et al., 1989).

The first reversible optical storage based on a photochromic reaction in polymeric liquid crystals was reported by Eich et al. (1987a). The authors reported erasable holographic recording in the liquid crystal polymers, containing cyano-substituted azobenzene side groups (Figure 4.17). These groups undergo reversible trans–cis isomerization on irradiation with suitable light, resulting in a change of the chromophore conformation. This was proved by the measurements of polarization spectra changes during cis–trans isomerization of the azobenzene (Figure 4.23). This in turn brings about change of the order parameter S (see eqn. 3).

In the optically clear monodomain of the liquid crystal polymer (oriented homeotropically or homogeneously) the photoinduced trans–cis isomerization of the azobenzene groups produced reorientation of the surrounding mesogenic groups and induced birefringence $\Delta n = 7.1^{-3}$. At room temperature and with green ($\lambda = 514.5$ nm) light of intensity approximately 10 mW/cm^2 this birefringence change could be obtained after 10 s of irradiation. This exceeded by two orders of magnitude the sensitivity achieved on thermo-recording. Note that the photochemical mesophase reorientation can also be considered as a non-linear, collective response of mesogenic groups to a steric change of a single photoactive azobenzene group. The recorded signal could be easily erased by heating above T_g. It is remarkable that spontaneous *cis–trans* thermal conversion which proceeds rather fast at room temperature does not lead to a vanishing of the recorded information.

The method was proved on different holographic techniques, and an erasable hologram of a real object on the polymeric material was demonstrated for the first time (Eich *et al.*, 1987b).

Orlter *et al.* (1989) and co-workers also incorporated azobenzene units in the cholesteric polysiloxane depicted in Figure 4.38 (R^3 and R^4). This enabled them to carry out efficient holographic recording based on the photoinduced orientation of the cholesteric films previously aligned parallel to the glass surface. Erasure of the recorded hologram could be easily performed by heating above T_g. A good hologram efficiency was achieved though the sensitivity was rather low. Three hundred recording–erasure cycles were performed without considerable change in efficiency.

In conclusion, the photochromic liquid crystal polymers offer a unique combination of structural, optical and electrical properties. Their synthesis and investigations were started only a few years ago, and the number of reported systems is still very limited. Many obvious structural variations have not yet been realized, e.g. incorporation of photochromic groups into a main chain or binding of photochromic side groups to main-chain liquid crystal polymers. Also, the types of photochromes studied are limited only to spiro and azo derivatives. However, the author believes that the new linear and especially non-linear effects, such as the rheo-optical effect, SHG or the photoinduced change of mesophase arrangement, disclosed during this short period warrant further research into these fascinating materials.

Acknowledgements

The author wishes to thank Professor Ernst Fischer for stimulating discussion and The German–Israeli Foundation for Scientific Research for financial support.

References

Asada, T., Muramatsu, H. Watanabe, R. and Onogi, S. (1980) *Macromolecules* **13**, 867.
Attard, G. and Williams, G. (1987) *Nature (Lond.)* **326**, 544.
Bergmann, E., Weizmann, A. and Fischer, E. (1950) *J. Am. Chem. Soc.* **72**, 5009.
Bird, R., Armstrong, R. and Hassager, O. (1977) *Dynamics of Polymeric Liquids*, Vol. 1. Wiley, New York, Chap. 4.
Bloembergen, N. (1965) *Nonlinear Optics*. Benjamin, New York, p. 69.
Blumstein, A. (ed.) (1985) *Polymeric Liquid Crystals*. Plenum Press, New York.
Boyd, G.T. (1989) *J. Opt. Soc. Am.* **B.6**, 685.
Cabrera, I. and Krongauz, V. (1987a) *Nature (Lond.)* **326**, 582.
Cabrera, I. and Krongauz, V. (1987b) *Macromolecules* **20**, 2713.
Cabrera, I., Shvartsman, F., Veinberg, O. and Krongauz, V. (1984) *Science* **226**, 341.

172 APPLIED PHOTOCHROMIC POLYMER SYSTEMS

Cabrera, I., Krongauz, V. and Ringsdorf, H. (1987) *Angew. Chem. Int. Ed. Engl.* **26**, 1178.
Chandrasekhar, S. (1977) *Liquid Crystals*. Cambridge University Press, Cambridge, 83.
Chapoy, L.L. (ed.) (1985) *Recent Advances in Liquid Crystalline Polymers*. Elsevier, London.
Coles, H.J. and Simon, R. (1985) *Polymer* **26**, 1801.
De Jeu, W.H. (1980) *Physical Properties of Liquid Crystalline Materials*. Gordon & Breach, New York.
Durbin, S. and Shen, Y.R. (1984) *Phys. Rev.* **A30**, 1419.
Eckhardt, H., Bose, A. and Krongauz, V. (1987) *Polymer* **28**, 1959.
Eich, M. and Wendorff, J.H. (1987a) *Makromol. Chem. Rapid Commun.* **8**, 967.
Eich, M., Wendorff, J.H., Reck, B. and Ringsdorf, H. (1987b) *Makromol. Chem. Rapid Commun.* **8**, 59.
Eidenschink, R. (1989) *Angew. Chem. Int. Ed. Adv. Mater.* **28**, 1424.
Eisenbach, C.D. (1978) *Makromol. Chem.* **179**, 2489
Eisenbach, C.D. (1980) *Ber Bunsenges. Phys. Chem.* **84**, 680.
Eisenbach, C.D. (1980) *Makromol. Chem.* **179**, 2489.
Finkelmann, H. (1987) *Angew. Chemie Int. Ed. Engl.* **26**, 816.
Finkelmann, H., Kook, H. and Rehage, G. (1984) *Makromol. Chem. Rapid Commun.* **5**, 287.
Goldburt, E., Shvartsman, F., Fishman, S. and Krongauz, V. (1984a) *Macromolecules* **17**, 1225.
Goldburt, E., Shvartsman, F. and Krongauz, V. (1984b) *Macromolecules* **17**, 1976.
Gray, G.W. (1989) In *Side Chain Liquid Crystal Polymers*, ed. Mc Ardle, C.B. Blackie and Son, Glasgow, p. 106.
Hann, R.A. and Bloor, D. (eds) (1989) *Organic Materials for Non-Linear Optics*. The Royal Society of Chemistry, London.
Horie, K., Tsukamoto, M. and Mita, I. (1985) *Eur. Polym. J.* **21**, 805.
Hsiung, H., Rasing, Th. Shen, Y.R. Shvartsman, F., Cabrera, I. and Krongauz, V. (1987) *J. Chem. Phys.* **87**, 3127.
Kelker, H. and Hatz, R. (1980) *Handbook of Liquid Crystals*. Verlag Chemie, Weinheim.
Krongauz, V.A. (1979) *Israel J. Chem.* **18**, 304.
Krongauz, V. (1990) In *Photochromism — Molecules and Systems*, eds. Duerr, H. and Bouas-Laurent, H. Elsevier, Amsterdam (in press).
Krongauz, V.A. and Goldburt, E.S. (1981) *Macromolecules* **14**, 1382.
Krongauz, V.A., Fishman, S.N. and Goldburt, E.S. (1978) *J. Phys. Chem.* **82**, 2469.
Kryszewski, M. and Nadolski, B. (1977) *Pure Appl. Chem.* **49**, 511.
Kumar, G.S. and Neckers, D.C. (1989) *Chem. Rev.* **89**, 1917.
Kuzma, M. (1983) *Mol. Cryst. Liq. Cryst.* **101**, 351.
Lylel, R., Lipscomb, G.F., Thackara, J., Altman, J., Elizondo, P., Stiller, M. and Sullivan, B. (1987) *SPIE Conf. Proc.* **824**, 152.
Lyons, M.H. (ed.) (1989) *Materials for Non-Linear and Electro-Optics*, Institute of Physics Conference Series N103.
McArdle, C.B. (ed.) (1989) *Side Chain Liquid Crystal Polymers*, Blackie and Son, Glasgow.
McArdle, C.B., Clark, M.G., Haws, C.M., Wiltshire, M.C.K., Parker, A., Nestor, G., Gray, G.W., Lacey D. and Toyne, K.J. (1986) *Liq. Cryst.* **1**, 281.
McRae, E.D. and Kasha, M. (1964) *Physical Processes in Radiation Biology*, Academic Press, New York, 23.
Meirovitch, E., Shvartsman, F.P. and Krongauz, V.A. (1985) *J. Phys. Chem.* **89**, 5522.
Meredith, G.R., VanDusen, J.G. and Williams, D.J. (1982) *Macromolecules* **15**, 1385.
Ortler, R., Bräuchle, Chr., Miller, A. and Riipl, G. (1989) *Makromol. Chem. Rapid Commun.* **10**, 189.
Parshutkin, A.A. and Krongauz, V.A. (1974) *Molecular Photochem.* **6**, 437.
Patel, J., Leslie, T. and Goodby, J.W. (1984) *J. Ferroelectrics* **59**, 137.
Pinsl, J., Bräuchle, Chr. and Kruezer, F.H., (1987) *J. Mol. Electronics* **3**, 9.
Portugal, H., Ringsdorf, H. and Zentel, R. (1982) *Makromol. Chem.* **183**, 2311.
Prasad, P.N. and Ulrich, D.R. (eds) (1988) *Non-Linear Optical and Electroactive Polymers*. Plenum Press, New York.
Priestly, E., Wojtowicz, P. and Sheng (1975) *Introduction to Liquid Crystals*. Plenum Press, New York, Chap. 4.
Richert, R. and Bassler, H. (1985) *Chem. Phys. Lett.* **116**, 302.
Ringsdorf, H. and Schmidt, H.W. (1984) *Makromol. Chem.* **185**, 1327.
Ringsdorf, H. and Zentel, R. (1982) *Makromol. Chem.* **183**, 1245.

Ringsdorf, H., Schmidt, H.W. and Schneller, A. (1982) *Makromol. Chem. Rapid Commun.* **3**, 745.

Ringsdorf, H., Schmidt, H.W., Baur, G. and Keifer, R. (1985) In *Recent Advances in Liquid Crystalline Polymers*, ed. Chapoy, L.L., Elsevier, London, 253.

Ringsdorf, H., Schmidt, H.W., Eilingsfeld, H. and Etzback, K.H. (1988) *Makromol. Chem.* **188**, 1355.

Ringsdorf, H., Schrlab, B. and Venzmer, J. (1988b), *Angew. Chemie. Int. Ed. Engl.* **27**, 114.

Schmidt, H.W. (1989) *Angew. Chem. Int. Ed. Engl. Adv. Mater.* **28**, 940.

Schneider, S.Z. (1987) *Z. phys. Chem.* **154**, 91.

Shen, Y.R. (1984) *The Principles of Nonlinear Optics*. Wiley, New York, Chaps. 2 and 7.

Shibaev, V.P., Kostromin, S.G., Plate, N.A., Ivanov, S.A., Vetrov, V.Y. and Yakovlev, I.A. (1985) In *Polymeric Liquid Crystals*, ed. Blumstein, A. Plenum Press, New York, p. 345.

Shragina, L., Buchholtz, F., Yitzchaik, S. and Krongauz, V. (1990) *Liq. Cryst.* **7**, 643.

Shvartsman, F. and Krongauz, V. (1984a) *J. Phys. Chem.* **88**, 6448.

Shvartsman, F. and Krongauz, V. (1984b) *Nature (Lond.)* 309, 608.

Shvartsman, F., Cabrera, I., Weis, A., Wachtel, E. and Krongauz, V.A. (1985) *J. Phys. Chem.* **89**, 3941.

Singer, K.D., Sohn, J.E. and Lalama, S.J. (1986) *Appl. Phys. Lett.* **49**, 248.

Smets, G. (1972) *Pure Appl. Chem.* **30**, 1.

Sturmer, D.M. and Heseltine, D.W. (1977) In *The Theory of the Photographic Process*, ed. James, H. Macmillan, New York, Chaps. 7 and 8.

Vertogen, G. and De Jeu, W.H. (1988) *Thermotropic Liquid Crystals, Fundamentals*. Springer Verlag, Berlin.

Wendorff, J.H. and Eich, M. (1989) *Mol. Cryst. Liq. Cryst.* **169**, 133.

Williams, D.J. (ed.) (1983) *Non-Linear Optical Properties of Organic and Polymeric Materials*, ACS Symposium Series, N233. American Chemical Society, Washington DC.

Williams, D.J. (1984) *Angew. Chem. Int. Ed. Engl.* **23**, 690.

Wismontski-Knittel, T. and Krongauz, V. (1985) *Macromolecules* **18**, 2124.

Yitzchaik, S., Ratner, J., Buchholtz, F. and Krongauz, V. (1990a) *Liquid Cryst.*, in press.

Yitzchaik, S., Cabrera, I., Buchholtz, F. and Krongauz, V. (1990b) *Macromolecules* **23**, 707.

Yitzchaik, S., Berkovic, G. and Krongauz, V. (1990c) *Chem. Mater.* **2**, 162.

Yitzchaik, S., Berkovic, G. and Krongauz, V. (1990d) *Adv. Mater.* **2**, 33.

Yitzchaik, S., Berkovic, G. and Krongauz, V. (1990e) *Macromolecules* **23**, 3539.

Yitzchaik, S., Berkovic, G. and Krongauz, V. (1990f) *Optics Lett.*, in press.

Zajtseva, E., Prohoda, A., Kurkovskaya, L., Shifrina, R., Kardash, N., Drabkina, D. and Krongauz, V. (1973) *Chem. Heterocyclic Comp.* **9**, 1233.

Zyss, J. (1985) *J. Mol. Elect.* **1**, 25.

5 Photoresponsive polymers: reversible control of polymer conformation in solution and gel phases

M. IRIE

5.1 Introduction

In this chapter attention is focused on the photostimulated conformation change of polymers, because conformation is one of the most important fundamental properties of polymers and is relevant as a general model of photoresponsive polymers. The conformation of polymers governs their various physicochemical properties. When the conformation is reversibly controlled by photoirradiation, the conformation change should produce a concomitant change in polymer properties in solution and in gel phases. Polymers which change their properties reversibly by photoirradiation are named photoresponsive polymers. This chapter describes several attempts to construct such photoresponsive polymers by way of incorporated photochromic chromophores. For completeness, it should be noted that several other properties of photoresponsive polymers which do not depend on polymer conformation can be regulated; these are listed in Table 5.1. In such applications, photochromic reactions are used to trigger polymers, thereby imposing control of their properties with a view to accomplishing specific functions.

Table 5.1 Physical and chemical properties of photoresponsive polymers controlled by photoirradiation.

Solution	Solid
Viscosity	Membrane potential
pH	Membrane permeability
Solubility	Surface wettability
Metal ion capture	Shape
Capability	Sol–gel transition
	Miscibility of polymer blend

The polymer chain conformation is known to depend on the environment, such as solvent or temperature. In good solvents, polymers have an extended conformation, while they shrink in poor solvents at low temperature. Polyelectrolytes change their conformation with changes in pH and salt concentration. It is obviously tedious to change the environment in order to control the chain conformation. If the conformation could be controlled reversibly by photoirradiation outside the sample, the method of control would be superior to the above method in terms of the response time, reversibility and ease of procedure. Photochromic chromophores are useful to control the conformation by photoirradiation when they are incorporated into the polymer backbones or pendant groups.

5.2 Photostimulated conformation changes of polymers in solution

Photoresponsive polymers consist of a photoreceptor part, which contains photochromic molecules, and a functional part (Figure 5.1; Irie, 1990). At first, an optical signal is captured by the photochromic chromophores and is converted to a chemical signal owing to the induced photoisomerization. Then, the chemical signal is transferred to the functional part via a chemical circuit, and this in turn controls the polymer properties. In some cases, amplification mechanisms are involved. To begin with, the guiding principle for designing polymers which change the conformation reversibly by photoirradiation will be outlined.

Figure 5.1 The structure of a photoresponsive polymer.

Figure 5.2 illustrates several ways of using photochromic chromophores as a tool for conformation changes. The first mechanism (Figure 5.2, 1) was proposed for the first time by Lovrien (1967). If a polymer is in equilibrium interaction with some photochromic low-molecular-weight chromophores, it may undergo a conformation change when irradiated with light, because the interaction between the polymer and the chromophores changes. The example described by Lovrien is the mixture of poly (methacrylic acid) and chrysophenine G (**1**, CHP) in water. CHP changes the hydrophobic property when the configuration changes from all *trans* to the *cis-trans-cis* (*c-t-c*) form.

Upon UV irradiation, *trans*-CHP isomerizes to the *cis* form (around 10%), and the aqueous solution viscosity decreases by as much as 80%. The conformation change was interpreted as follows. The anionic linear and planar all-*trans* CHP would attach itself to the hydrophobic poly(methacrylic acid)

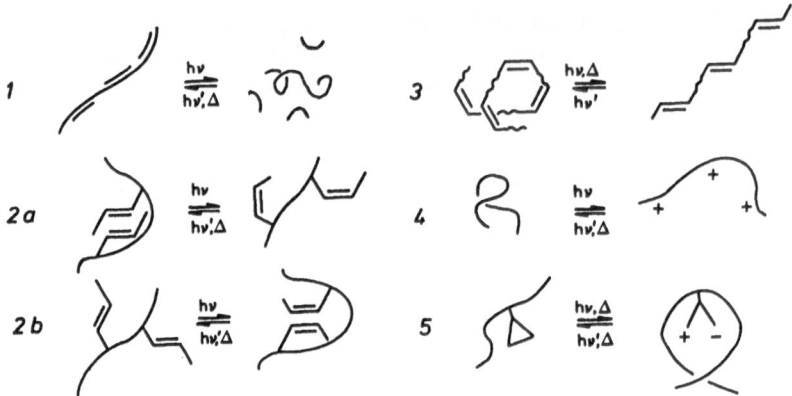

Figure 5.2 Schematic illustration of photostimulated conformation changes of polymer chains.

1

backbone, leading to an extended polymer conformation. *Cis* form azo dyes are, however, much more hydrophilic. Consequently, the *cis* form was envisaged as binding less strongly so that the polymer chain would be less extended.

The above system was re-examined by Van der Veen and Prins (1974). They, however, did not observe as large a change in viscosity as did Lovrien. The reduced viscosity of 43% for *cis*-CHP was only 5% lower than the value for CHP completely converted to the *trans* form. It was concluded that the viscosity decrease reported by Lovrien had been caused by some impurity contained in the dye and/or the polymer.

Efforts to confirm the photoregulation mechanism postulated by Lovrien have continued. Negishi *et al.* (1977) found a pronounced photostimulated viscosity change of an aqueous solution of 2-hydroxyethyl methacrylate (HEMA)–*N*-vinylpyrrolidone (VPy) copolymer and CHP or acid yellow 38 (**2**). The reduced viscosity reversibly decreased by as much as 12%. The interaction between the polymer and CHP was successfully controlled by photoirradiation. The result suggested that the mechanism should strongly depend on the pair of polymer and dyes (Negishi *et al.*, 1982).

The second mechanism (Figure 5.2, 2) utilizes the change induced in the intramolecular interaction between pendant groups by photoirradiation. The

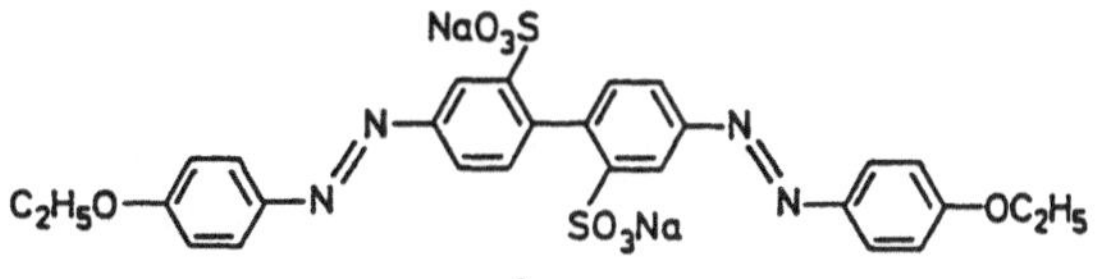

2

system reported for the first time is poly (methacrylic acid) with pendant azobenzene groups (Lovrien *et al.*, 1967). In an aqueous solution, the viscosity was found to be increased by UV irradiation (Figure 5.2, 2a). The *trans* to *cis* isomerization was considered to decrease the hydrophobic interaction between the azobenzene chromophores, allowing the polymer coil to expand.

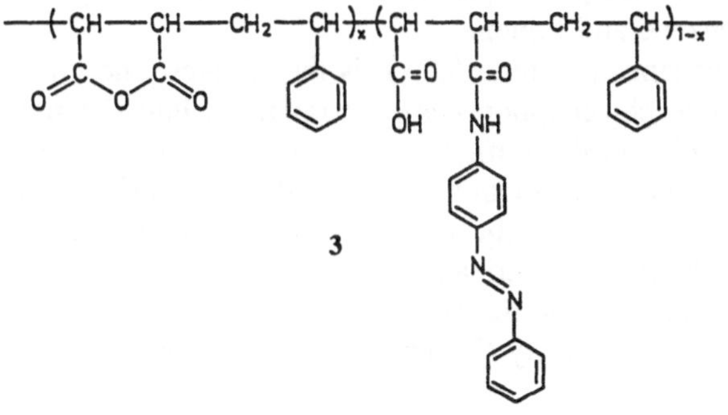

3

Matějka and Dušek (1981) have extended the study to a styrene–maleic anhydride copolymer with pendant azobenzene groups (**3**) and measured the photoresponsive behaviour in less polar solvents. This copolymer exhibited a reversible photodecrease of the viscosity in 1,4-dioxane solution. A decrease of 24–30% in the reduced viscosity was found after the solution was irradiated with UV light. In tetrahydrofuran, the viscosity decrease was 1–8%. The contraction of the dimensions of the copolymer coil was explained as follows. *Cis*-form azobenzene has a dipole moment of 3.1 D, while the dipole moment of the *trans* form is less than 0.5 D. Therefore, the *trans* to *cis* isomerization induces strong dipoles in the pendant groups. These dipoles tend to orient in parallel and attract each other in less polar solvents, so that compact coil conformations are preferred, as shown in Figure 5.2 (2b). In the dark, the viscosity of the copolymer solution returned to the original value, though the process occurred much more slowly. The rate was 1/2.5 to 1/7 of the rate of *cis* to *trans* isomerization of the pendant azobenzene chromophores, which was measured by optical absorption. The discrepancy requires further examination of the postulated mechanism for the conformation change.

The dipole moment increase of azobenzene residues by photoirradiation can also induce a change in polymer chain conformation even when the azobenzene chromophores are incorporated into the polymer backbone The solution viscosity of poly (dimethylsiloxane) containing azobenzene residues in the main chain decreased upon UV irradiation, and the effect was attributed to the *trans* to *cis* photoisomerization (Irie and Suzuki, 1987). The photodecrease of the viscosity depended on the polarity of the solvent. It was 24% in non-polar heptane but negligible in polar dichloroethane. Stacking of the *cis*-form azobenzenes in less polar solvent is responsible for the conformation change.

Aggregate formation of merocyanine dyes is also useful to control the conformation. This system was developed by Goldburt *et al.* (1984) using spirobenzopyrans as the pendant photochromic molecules. They observed polymer stacking owing to the aggregation of photogenerated pendant merocyanine forms for poly (spiropyran methacrylate) and poly (spiropyran acrylate) in benzene and toluene.

The third mechanism (Figure 5.2, 3) is the simplest one. When *trans–cis* photoisomerizable chromophores are incorporated into the polymer backbone, the photoinduced configuration change of the chromophores is expected to induce a conformation change of the polymer chain. Azobenzene is the most widely used as the *trans–cis* photoisomerizable photoreceptor molecule. It undergoes isomerization from the *trans* (**4**) to the *cis* (**5**) form under UV irradiation, while the *cis* form can return to the *trans* form either thermally or photochemically (Zimmerman *et al.*, 1985).

During the course of isomerization, azobenzene undergoes a large structural change. The distance between the 4 and 4^1 carbons decreases from 9.0 to 5.5 Å (Hartley, 1938; Hampson and Robertson, 1941). Figure 5.3

Structure

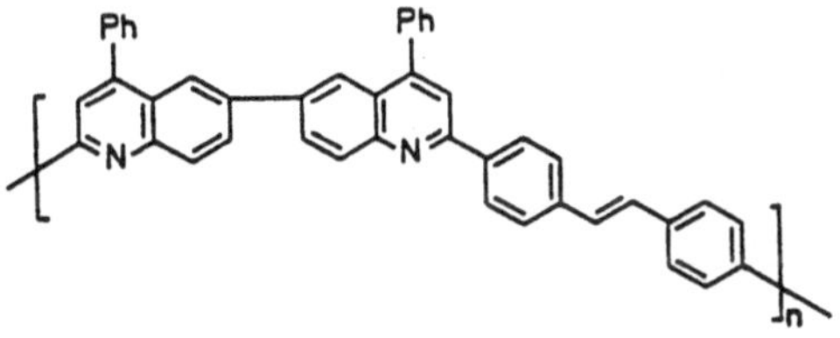

Figure 5.3 Photoresponsive polymers with photochromic chromophores in the backbone.

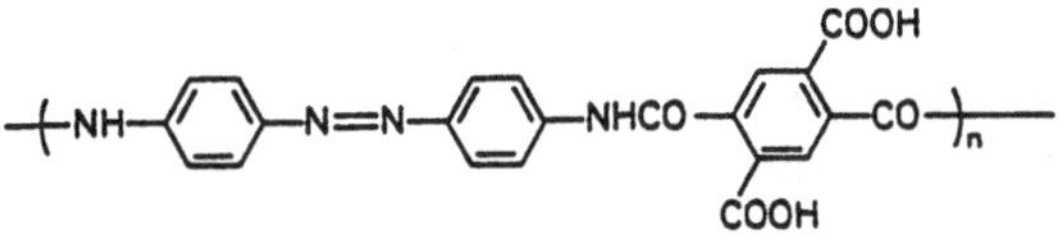

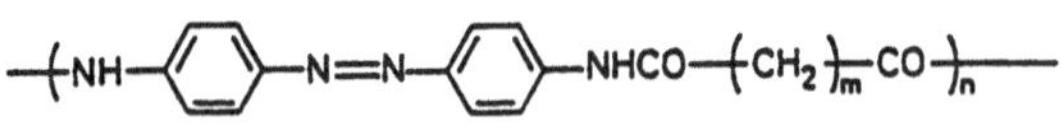

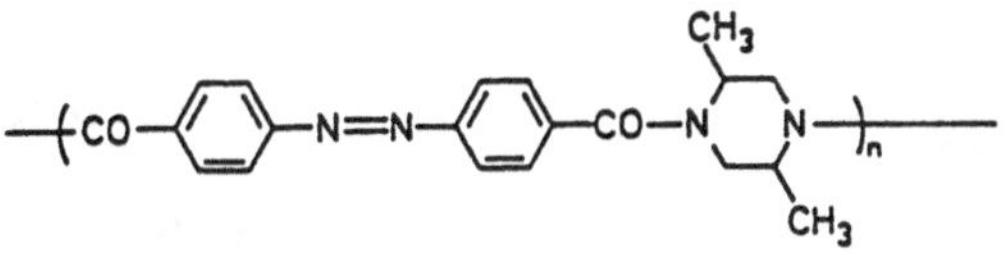

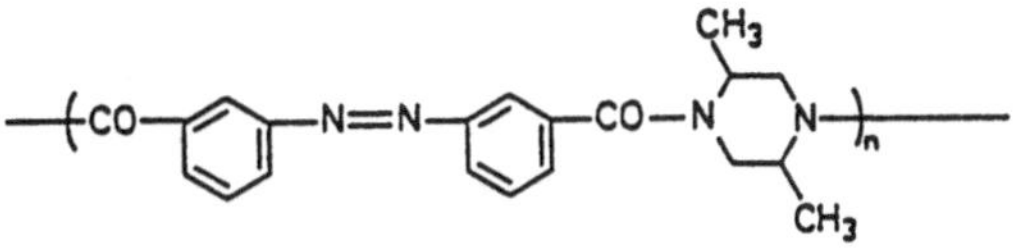

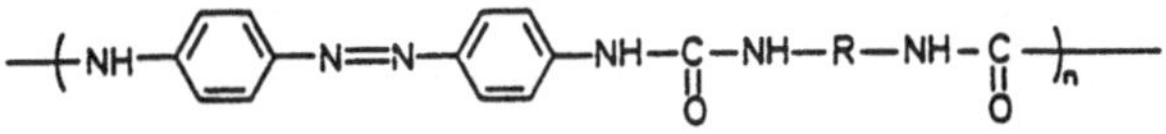

summarizes the reported polymers having photoisomerizable unsaturated linkages in their backbones, mostly containing azobenzene groups.

Polyamides with azobenzene groups in the backbone are among the earliest in which *trans–cis* isomerizable chromophores were used to regulate the polymer conformation (Irie and Hayashi, 1979; Irie *et al.*, 1981a). The intrinsic viscosity of polyamide (**6**) in polar *N,N*-dimethylacetamide was found to decrease from 1.22 to 0.5 dl/g upon UV irradiation (410 > λ > 350 nm) and to return to the initial value after 30 h in the dark at 20°C. The slow recovery of the viscosity in the dark was accelerated by visible light irradiation (λ > 470 nm). On alternate irradiation with UV and visible light, the viscosity changed as much as 60%, as shown in Figure 5.4. Before photoirradiation, the polyamide has a rod-like conformation. The isomerization from the *trans* to the *cis* form kinks the polymer chain, resulting in a compact conformation and a decrease in the viscosity. The compact conformation returns to the initial extended conformation either thermally or with visible light irradiation, thereby causing an increase in the viscosity.

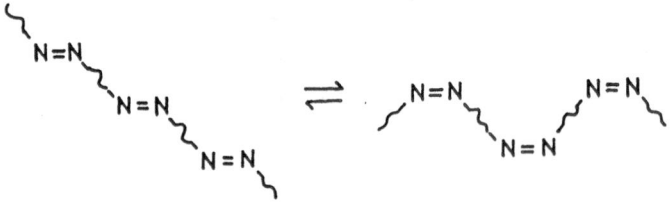

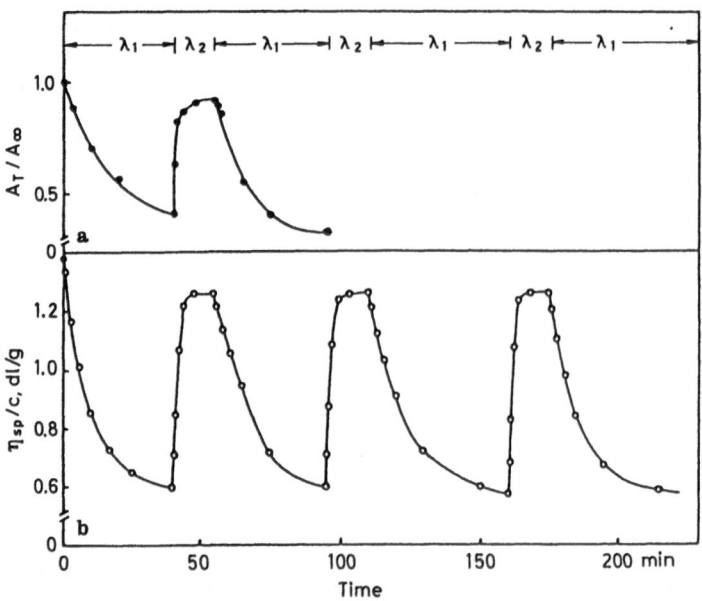

Figure 5.4 Changes in (•) content of the *trans*-azobenzene residues in polyamide (structure 6) backbone and (o) viscosity of the polyamide in *N,N*-dimethylacetamide on alternate irradiation with ultraviolet (410 > λ > 350 nm) and visible (λ > 470 nm) light at 20°C. Polymer concentration was 0.9 g/dl.

Table 5.2 Effect of backbone structure of the polymers on the photodecrease of the solution viscosity (cf. Figure 5.3).

Polymer	R^b	η_{sp} (UV)/ η_{sp} (dark)^a
Polyamide	$(p-C_6H_4)$	0.37
Polyamide 7	$(CH_2)_4$	0.59
Polyamide 7	$(CH_2)_8$	0.80
Polyamide 7	$(CH_2)_{12}$	0.96

$^a\eta_{sp}$ (UV) and η_{sp} (dark) are specific viscosities under irradiation with UV light (410 > λ > 350 nm) and in the dark before irradiation in N-methyl-2-pyrrolidone in the presence of lithium chloride (1.3 M) respectively.
^{b}R group in (NH — Ph — N=N — Ph — NHCO–R–CO)$_n$

The changing conformation of structure **6** illustrates that rigidity of the polymer chain is expected to alter the amount of photodecrease of the solution viscosity. When the azobenzene residues are connected by rigid phenylene groups, the resulting viscosity change should be large, whereas it should become small when the connecting groups are flexible, e.g. are long methylene chains. The effect of backbone structure on the photodecrease of solution viscosity is summarized in Table 5.2. As expected, the amount of photodecrease of the viscosity decreases with the increasing number of methylene groups in the polymer backbone. The stiffest polymer (that having phenylene residues) gives a large photodecrease, while the viscosity of polymers having flexible long methylene chains is hardly reduced by photoirradiation. The absence of photodecrease in the polymer with 12 methylene groups suggests that flexible methylene chains act as a strain absorber. The conformation change induced by the isomerization of the azobenzene residues is relaxed in the connecting flexible methylene chains, resulting in no change of the shape of the polymer.

Similar experiments were carried out for polyamides (**8** and **9**) by Blair et al. (1980) but they found only a small decrease in the reduced viscosity at a high polymer concentration. The absence of photodecrease in the intrinsic viscosity is probably because of the inclusion of flexible piperazine segments in the polymer chain.

Neckers and co-workers (Kumar et al. 1984a, b, 1985) demonstrated that polyureas with backbone azobenzene groups (**10**) also underwent a photoviscosity effect when UV irradiated. Zimmerman and Stille (1985) reported that the intrinsic viscosity of polyquinoline (**11**) with backbone stilbene groups in di-m-cresyl phosphate/m-cresol decreased as much as 24% under UV light. The decrease was ascribed to the *trans* to *cis* isomerization of the stilbene groups. Because of its simplicity the mechanism (Figure 5.2, 3) has been widely applied to polycondensation or polyaddition polymers.

The fourth mechanism (Figure 5.2, 4) employs the electrostatic force of repulsion between photogenerated charges as the driving force for conformation changes. Triphenylmethane leuco derivatives have been used as photoreceptor molecules. The chromophore dissociates into an ion pair under UV irradiation, generating an intensely green-colored triphenylmethyl cation. The cation thermally recombines with the counteranion as follows:

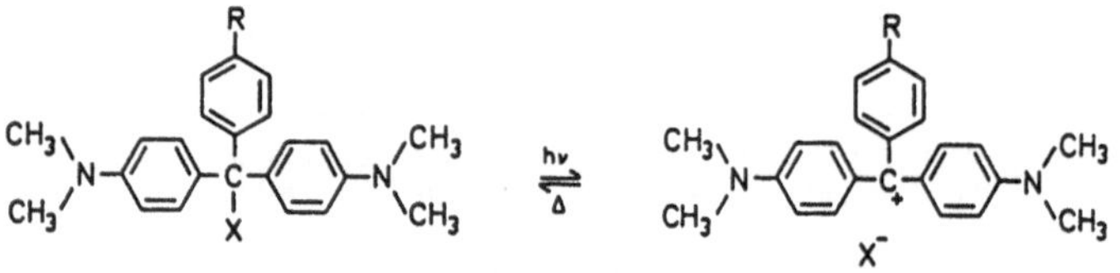

12

13

The triphenylmethane leucohydroxide residues were incorporated into the pendant groups by copolymerizing the vinyl derivative (**12**, R = CH=CH$_2$, X=OH) with *N,N*-dimethylacrylamide (Irie and Hosoda, 1985). Upon irradiation ($\lambda > 270$ nm), the solution became deep-green, and at the same time its reduced viscosity increased from 0.55 to 1.6 dl/g, as depicted in Figure 5.5. After removal of the light the viscosity as well as the absorption returned to the initial value with a half-life of 3.1 min. The close correlation between the viscosity change and the absorption intensity at 620 nm implies that the electrostatic repulsion was responsible for the expansion of the polymer conformation.

The concentration dependence of reduced viscosity confirmed the above expansion mechanism. In the dark before photoirradiation, the dependence was linear; the reduced viscosity decreased with decreasing polymer concentration. During UV irradiation, this dependence showed anomalous behaviour. The viscosity steeply increased at low polymer concentration. The viscosity during photoirradiation was four times larger than the viscosity in the dark at 0.04 g/dl. The concentration dependence is characteristic of polyelectrolytes. At low polymer concentration, screening of the electrostatic potential by the counterions becomes weak, and consequently the increase of the repulsive forces of positive charges along the polymer chain expands the dimension of the chain. The photostimulated increase of the viscosity was suppressed in the presence of salt (10^{-3} *M* lithiumbromide). The salt effect is definite evidence of the above-mentioned mechanism. The photoeffect due to the electrostatic forces is much larger than the effect observed for polyamides with azobenzene residues in the backbone.

Formation of strong dipoles along the polymer chain and the intramolecular interaction of the dipoles with the polymer chain would also change the chain conformation, as expected by the fifth mechanism (Figure 5.2, 5). This approach to induce conformation changes has used spirobenzopyran or azobenzene groups as photoreceptor molecules. It is well known

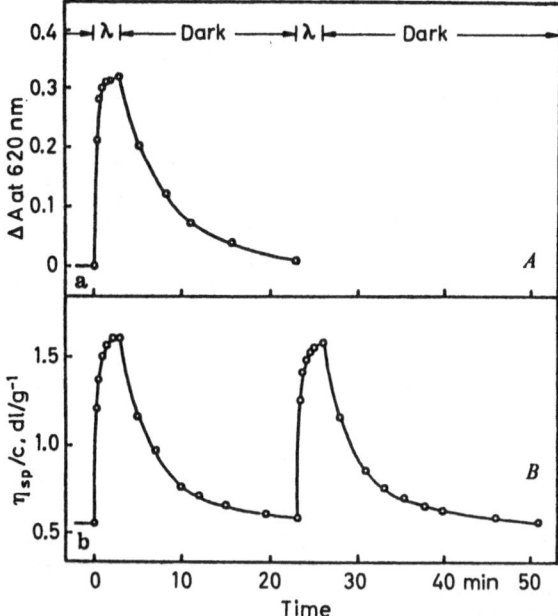

Figure 5.5 Changes of (A) absorption at 620 nm and (B) viscosity of poly(*N,N*-dimethylacrylamide) having pendant triphenylmethane leucohydroxide groups [9.1% (mol/mol)] in methanol at 30°C on exposure to UV light (λ > 270 nm). Polymer concentration was 0.06 g/dl.

that spirobenzopyran (**14**) under UV light irradiation undergoes a ring-opening reaction leading to the formation of merocyanine (**15**) which has a strong dipole. This reaction can be reversed either thermally or photochemically.

14   **15**

The dipole moment change can be used as a driving force for conformation changes of the polymer chain. This is done by incorporating the spirobenzopyrans into the pendant groups. A typical example is poly (methyl methacrylate) with pendant spirobenzopyran groups (Irie *et al.*, 1981b). Figure 5.6 shows its viscosity (the content of pendant spiropyran groups is 13%

(mol/mol) in benzene in the dark as well as during the photoirradiation ($\lambda > 310$ nm). The data for dichloroethane are also included to illustrate the viscosity behaviour in polar solvents. In benzene, the intrinsic viscosity during irradiation is 17% lower than the viscosity in the dark. The viscosity change in polar dichloroethane is only 1%. The effect of the polarity on the ratio of viscosity during photoirradiation to that in the dark was examined in several solvents and is shown in Table 5.3. The viscosity change decreases almost in parallel with increasing microscopic polarity.

Table 5.3 Solvent effect on the photostimulated viscosity change in spiropyranic polymethacrylates.

Solvent	$[\eta]_p/[\eta]_d$	$E_{T30}{}^a$
Benzene	0.83	34.1
Dioxane	0.87	36.1
Tetrahydrofuran	0.95	37.4
Ethylacetate	0.93	38.1
Chloroform	0.97	39.1
Dichloroethane	0.99	$(41.1)^b$

$^a E_{T30}$ value of Dimroth at 25°C.
^{b}Value of dichloromethane.

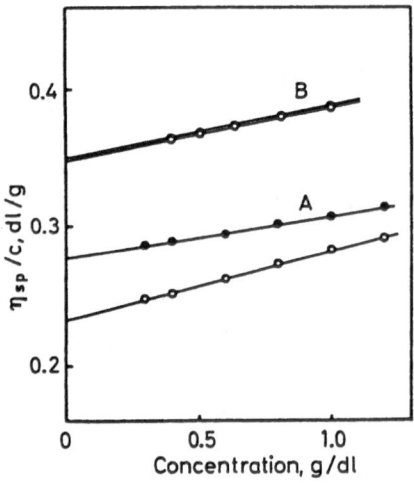

Figure 5.6 Viscosities of poly(methyl methacrylate) having spirobenzopyran groups [13% (mol/mol)] at 30°C: (A) in benzene (●) in the dark (o) under irradiation ($\lambda > 310$ nm); (B) in dichloroethane.

These solvent effects suggest that the polymer chain shrinks mainly as a result of specific solvation of the photogenerated merocyanines by the poly(methyl methacrylate) ester groups. This intramolecular solvation competes with the solvation by solvents. The intramolecular attraction between the pendant merocyanine and the polymer chain overcomes the merocyanine–solvents interaction, giving rise to a polymer with a more coiled conformation. The intramolecular dipole–dipole interaction between pendant merocyanine groups in a polymer is less likely to decrease viscosity for the following reasons: (1) the benzene viscosity of polystyrene having pendant spirobenzopyran groups showed no response to photoirradiation, and (2) the photodecrease of the benzene viscosity of the poly(methyl methacrylate) reached a maximum at as low a spirobenzopyran content as 17% (mol/mol).

Poly (methacrylic acid) with pendant spirobenzopyran groups also showed photostimulated conformation changes in methanol (Menju et al., 1981). Visible light irradiation of the solution increased the viscosity, while UV light irradiation caused the solution viscosity to decrease. Alternate irradiation with visible and UV light brought about reversible viscosity changes with an amplitude as large as 40%, and it was possible to repeat the cyclic changes in viscosity many times.

Photostimulated conformation changes observed for polypeptides with pendant azobenzene residues may also be classified according to mechanism 5 (Figure 5.2) (Ueno and Osa, 1980; Ciardelli et al., 1984).

Photochromic reactions have been successfully utilized to induce conformation changes. An optical signal is captured by the photochromic molecules and converts to chemical signals (geometrical structure and/or dipole moment changes) resulting from the isomerization of the chromophores. The chemical signals are transferred to the polymer chain, and eventually a change in polymer conformation results. This process is commonly observed for photoresponsive polymers, and this underlying principle can be applied for controlling various properties of polymers.

5.3   Photostimulated phase separation of polymer solutions — another approach to induce the conformation changes of polymers

The photoresponsive polymers described above change conformation in proportion to the number of photons that they absorb. When the number of photochromic chromophores that can undergo an isomerization by absorbing a definite number of photons depending on the quantum yield is increased in a polymer, the conformation changes more. Physical and chemical properties associated with the conformation changes also vary with the number of absorbed photons. To make a sensitive photoresponsive polymer, i.e. one which responds more efficiently to fewer photons, it is necessary to introduce an amplification mechanism to the system. A convenient way to achieve this end is to utilize the phase transition of polymer systems.

At a temperature close to the phase transition temperature, the system is in an unstable state, and hence a small perturbation may bring about a large effect. When such a system is perturbed by photochromic reactions of chromophores incorporated into the polymers, the absorption of a few photons will induce a large property change.

Figure 5.7 shows a schematic illustration of the photostimulated phase transition from the state X to the state Y. When the photochromic chromophores in the polymer chain are A isomers, the polymer changes its state at a temperature of T_a. It is assumed that the phase transition temperature will rise to T_b when A isomers convert to B isomers. Then, if the isomerization from the A to the B isomers can be induced by photoirradiation at temperature T ($T_a < T < T_b$), the state will change isothermally at T from Y to X by photoirradiation. When the states X and Y have different polymer conformation, a few photons can induce a marked change in the conformation as well as the physical and chemical properties. A typical phase transition which accompanies the conformation change is phase separation of polymer solutions. Two examples of this are described here.

Poly (N-isopropylacrylamide) in water has a low critical temperature of 31°C. The solution is homogeneous and the polymer has random conformation below 31°C, while it becomes inhomogeneous and the conformation shrinks above 31°C. The critical temperature T_c depends on the subtle balance between the ability of the polymer to form hydrogen bonds with water and the intermolecular hydrophobic forces. The hydrophobic interaction is expected to be controlled by photoirradiation when photoisomerizable chromophores are introduced into the polymer. Thus, the temperature at which this system undergoes phase separation is expected to be altered by photoirradiation. Such an attempt was made by incorporating azobenzene chromophores into the pendant groups (16) (Kungwatchakan and Irie, 1988).

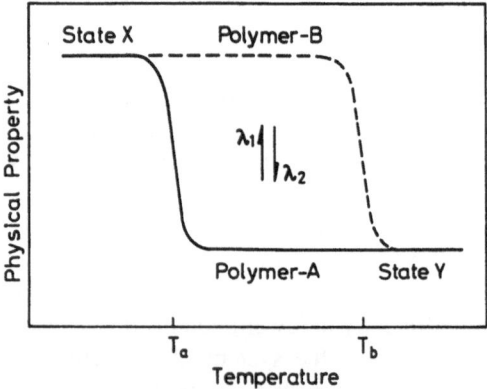

Figure 5.7 Schematic illustration of photostimulated phase transition from the state X to the state Y.

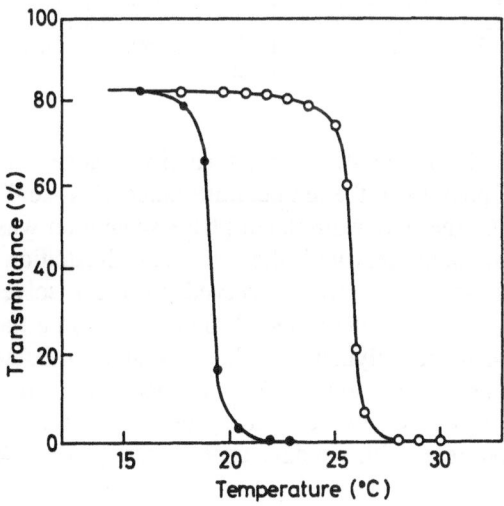

16

Figure 5.8 shows the transmittance change at 750 nm which occurred when a 1% aqueous solution of poly (*N*-isopropylacrylamide) with 2.7% (mol/mol) pendant azobenzene groups was heated. In the dark before photoirradiation, the solution begins to be turbid at 18.5°C and the transmittance decreases to one-half the initial value at 19.4°C. Upon UV irradiation (410 > λ > 350 nm), the phase separation temperature rises to

Figure 5.8 Transmittance changes at 750 nm of 1% aqueous solution of poly(*N*-isopropylacrylamide) with pendant azobenzene groups [2.7% (mol/mol)] when heated at a rate of 2°C/min. (•) Before photoirradiation; (o) under photostationary state with UV irradiation (410 > λ > 350 nm).

26.0°C. Between 19.4 and 26.0°C, UV irradiation solubilizes the polymer and the solution becomes transparent, while visible irradiation decreases the solubility of the polymer and leads to phase separation (see Figure 5.9).

The maximum difference in phase separation temperature was observed at a very small azobenzene content of 2.7% (mol/mol). Below and above this content, the phase separation was not affected by photoirradiation. This fact confirms that the phase transition temperature depends on a subtle balance between the polymer's ability to form hydrogen bonds with water and the intermolecular hydrophobic force. The isomerization of a small number of azobenzene chromophores [2–3% (mol/mol)] affects the balance, resulting in an efficient phase separation.

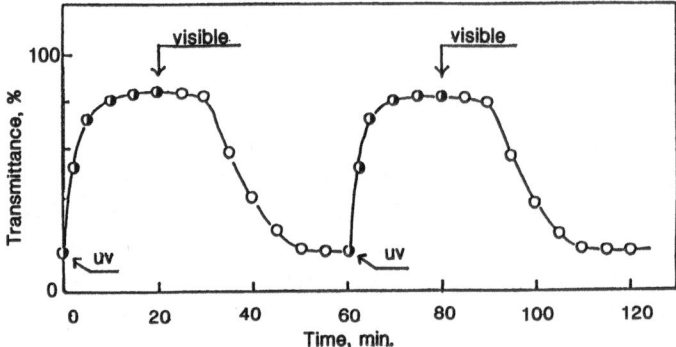

Figure 5.9 Reversible phase separation of aqueous solution of poly(*N*-isopropylacrylamide) with pendant azobenzene groups [2.7% (mol/mol)] at 19.5°C upon alternate irradiation with UV light (410 > λ > 350 nm) and visible light (λ > 450 nm). Transmittance was measured at 750 nm.

A large effect, in this case, a large turbidity change was induced by a small number of photons in the temperature range 19.4–26.0°C. Below 19.4 and above 26.0°C, the photostimulated phase separation was not observed. These findings are consistent with the schematic illustration in Figure 5.7.

Similar phase separation was observed in theta solvents containing polymers with pendant photochromic chromophores. In a theta solvent, the interaction between the polymer and the solvent is in balance with intra- and interpolymer interactions. The isomerization of the pendant chromophores alters this balance. The system studied was a cyclohexane solution of polystyrene with pendant azobenzene groups (see **17**; Irie and Tanaka, 1983).

Cyclohexane becomes a theta solvent for polystyrene at 35°C. Moderate-molecular-weight polystyrene (MW = 5 × 10⁴) with pendant azobenzene groups is soluble in this solvent at 30°C. The solution becomes turbid upon UV irradiation (410 > λ > 350 nm). Prolonged irradiation causes the polymer

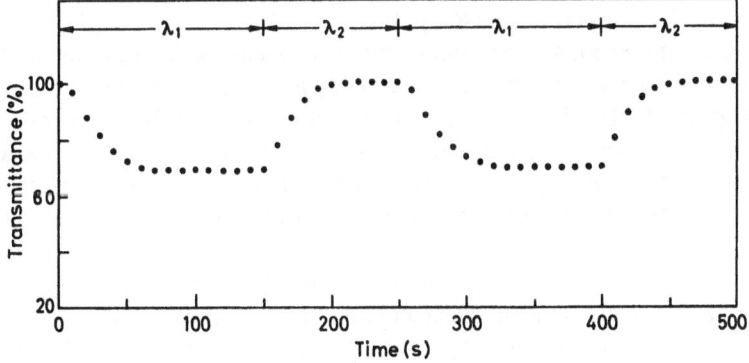

to precipitate. The solution becomes transparent when irradiated with visible light ($\lambda > 470$ nm). The photoresponsive behaviour is shown in Figure 5.10. The phase separation is due to the isomerization of the pendant azobenzene groups. Introduction of non-polar *trans*-form azobenzene chromophores into the pendant groups little affects the polymer–solvent interaction, while the photogenerated *cis* form tends to decrease the polymer–solvent interaction. Thus, upon UV irradiation, the polymer-solvent interaction decreases considerably until the polymer precipitates.

The precipitation behaviour of the polymer is interpreted by a photostimulated change of the critical miscibility temperature T_c. For polystyrene

Figure 5.10 Changes in transmittance at 650 nm of a cyclohexane solution containing polystyrene having pendant azobenzene groups [content 6.1% (mol/mol)] on alternate irradiation with UV ($410 > \lambda > 350$ nm) and visible ($\lambda > 470$ nm) light at 30°C.

dissolved in cyclohexane, the polymer precipitates at temperatures below T_c. According to Fox and Flory (1951), T_c depends on the molecular weight M as:

$$T_c = T_\theta (1 - b/M^{0.5})$$

where T_θ is the value of T_c for $M = \infty$ and b is an empirical constant. It is assumed that T_θ and b change when the *trans* azobenzene groups convert to the *cis* form. Figure 5.11 shows a schematic illustration of the molecular weight dependence of T_c, plotting T_c against $M^{-0.5}$.

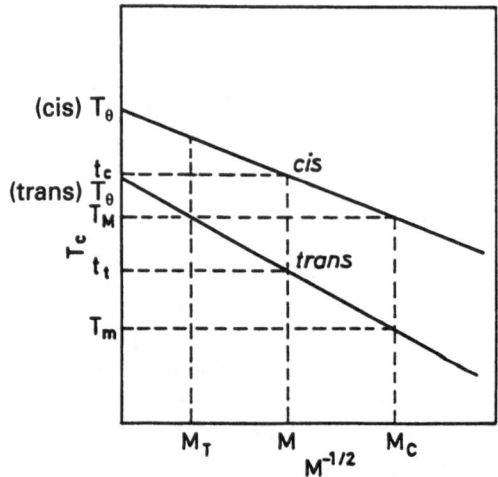

Figure 5.11 Molecular weight (M) dependence of critical miscible temperature, T_c. T_θ indicates the value of T_c for polystyrene with *cis-* and *trans*-azobenzene groups at $M = \infty$ respectively. See text for M, t_t, t_c, T_M and T_m.

First, consider a monodispersed polystyrene of molecular weight M containing pendant azobenzene groups. When azobenzene groups are in the *trans* form, the polymer solution phase separates at t_t, which corresponds to T_a of Figure 5.6. The isomerization of the chromophores from the *trans* to the *cis* form causes the phase separation temperature to rise to t_c, which corresponds to T_b of Figure 5.6. This means that phase separation of the solution is induced between t_t and t_c by UV irradiation, which causes the *trans–cis* isomerization.

Next, consider the phase separation of a polydispersed polystyrene with pendant azobenzene groups at constant temperature, T_M. With all azo groups in the *trans* configuration, any fractions of $M < M_T$ are soluble in cyclohexane at T_M, but after photoirradiation, the fractions of $M_c < M < M_T$ become insoluble and precipitate from the solution. Actually, irradiation here has the same effect as lowering the solution temperature from T_M to T_m.

During the course of the phase separation, the polymer chain initially shrinks and then phase separates from the solution. The dynamics of the coil contraction and the subsequent phase separation processes can be appropriately studied by time-resolved light-scattering measurements (Irie and Schnabi, 1985b).

The isomerization of photochromic chromophores can be induced in less than 10^{-8}s with a short laser pulse. The conformation change subsequent to it can be followed with a time-resolved light-scattering system combined with the short laser pulse source.

The Debye equation:

$$\frac{Kc}{R_\theta} = \frac{1}{\overline{M}_w} + \frac{16\pi^2 <s^2>}{3\lambda_0^2 \overline{M}_w} \sin^2(\theta/2) + 2A_2$$

relates the light-scattering intensity R_θ to the weight-average molecular weight M_w, the mean square radius of gyration $<s^2>$ and the second virial coefficient A_2 of a polymer in dilute solution. Here, $K = (2\ \pi^2 n^2/N_A\ \lambda_0^4)$ $(dn/dc)^2$, c is the polymer mass concentration, n_0 is the refractive index of the solvent, dn/dc is the specific refractive index increment, λ is the wavelength of the incident light and N_A is the Avogadro constant. Expansion of the polymer coil leads to an increase in $<s^2>$, which decreases R_θ.

Figure 5.12 shows oscillograms of the polystyrene with pendant azobenzene groups [4.3% (mol/mol)] in cyclohexane, demonstrating the decrease at 360 nm and the increase at 440 nm during and after the flash. As can been seen from the oscillograms, the change of the optical absorption occurs during the flash in less than 10^{-8} s.

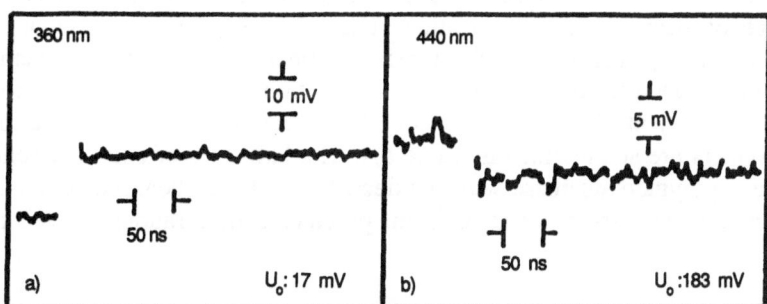

Figure 5.12 *Trans* to *cis* isomerization of pendant azobenzene groups after irradiation of cyclohexane solution of polystyrene with azobenzene groups [4.3% (mol/mol)]. Oscilloscope traces depicting the change of the optical absorption at (a) 360 and (b) 440 nm.

The change of the absorption is ascribed to *trans* → *cis* isomerization of the azobenzene groups. From the very fast *trans* → *cis* isomerization, it was inferred that neighbouring phenyl groups did not interact strongly with *trans*-azobenzene groups.

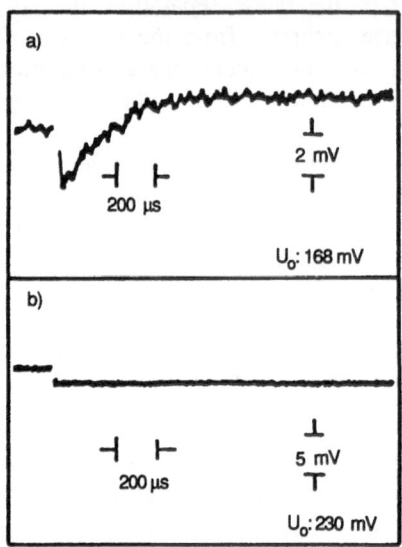

Figure 5.13 Oscillogram illustrating the increase of the light-scattering intensity (a) and the optical absorption (b), both at 514 nm, after irradiation of polystyrene with pendant azobenzene in groups in cyclohexane solution (0.13 g/l) with a laser of 347 nm light at 25°C.

In the microsecond time regime after the flash, an increase of the light-scattering intensity was observed. A typical oscillogram demonstrating the intensity change is shown in Figure 5.13. It can be seen that the intensity decreases initially. This decrease is due to the concurrent increase of the optical absorption at 514 nm (b) and of the decrease of dn/dc as a consequence of *trans* →*cis* isomerization. The relatively slow increase of the light-scattering intensity is considered to reflect the conformation change involving a decrease of both the radius of gyration of the coil and the second virial coefficient. The half-lives of the intensity increase for various copolymers are summarized in Table 5.4. The half-lives of coil contraction of the copolymers decrease with the decrease in the azobenzene content.

Experiments carried out at different polymer concentrations yielded the

Table 5.4 Conformational relaxation of polystyrene with pendant azobenzene groups.

Sample[b]	$M_w \times 10^{-4}$	$\tau_{1/2}(s)^a$
PS-A-4.3	2.7	2.1×10^{-4}
PS-A-5.6	2.0	4.0×10^{-4}
PS-A-6.5	1.8	8.5×10^{-4}

[a]Half-lives of the increase of the light-scattering intensity at 514 nm and 25°C.
[b]Polystyrene with pendant azobenzene groups. The number indicates the content of azobenzene groups [%(mol/mol)].

same half-life of the intensity change. This result is clear evidence that the process observed in the range of several hundred microseconds is an intramolecular reaction. The rate of coil contraction depended on temperature: $\tau_{1/2} = 4 \times 10^{-4}$s at 25°C and $\tau_{1/2} = 3 \times 10^{-4}$s at 35°C. Since the coil contraction depends on the mobility of both the segments and solvent molecules, it is expected to occur faster at higher temperature.

The fact that the isomerization of only a few azobenzene groups per chain induces significant conformation changes is attributed to a severe perturbation of the balance of polymer–solvent and polymer–polymer interactions by *trans* → *cis* isomerization of pendant azobenzene groups. The primary process of the phase separation is interpreted from the viewpoint of molecular mechanism as follows. Concurrent with the *trans* → *cis* isomerization, the dipole moment of pendant azobenzene groups increases from 0.5 to 3.1 D. It is feasible, therefore, that *cis*-azobenzene groups are more prone to interact with styrene base units than solvent cyclohexane molecules because of dipole-induced dipole interactions. *Trans*-azobenzene groups interact more favorably with solvent molecules than with styrene base units, whereas the reverse situation is true for *cis*-azobenzene groups, which interact less strongly with solvent molecules than with styrene base units. The enhanced capability of *cis*-azobenzene groups for interacting with other segments of the chain is considered to give rise to a shrinkage of the polymer chain.

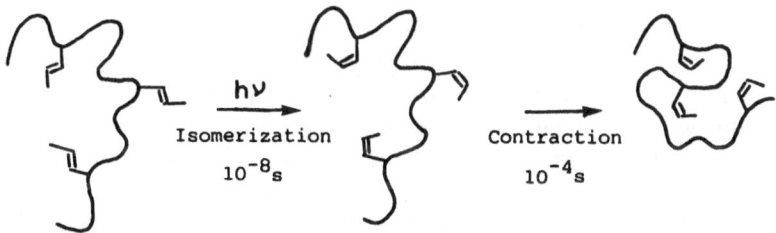

Figure 5.14 Schematic illustration of a conformation change of polystyrene with pendant azobenzene groups.

The process is schematically illustrated in Figure 5.14. The *trans* to *cis* isomerization of the pendant groups completes during the laser flash in less than 10^{-8}s. During the isomerization, the total chain conformation remains in the initial random conformation. The interaction between the *cis* azobenzene groups and styrene base units slowly changes the total chain conformation to a more stable contracted form in 0.1–0.8 ms. When different polymers collide with each other at a later stage, intermolecular interaction will also become operative. As a consequence, the polymer solution will undergo phase separation.

The photoresponsive behaviour described above is quite efficient. Only 5% (mol/mol) azobenzene pendant groups were enough to cause the conformation change. In the system containing spirobenzopyran groups, the efficiency was much higher (Irie et al., 1985). The isomerization of 2% (mol/mol) spirobenzopyran chromophores in the pendant groups raised T_c considerably and led to the conformation changes.

5.4   Photostimulated shape changes of polymer gels — continuous change

It is of particular interest to extend the photostimulated conformational change of polymer chains in solution to a macroscopic change in shape or size of gels or solids. The use of structural changes of photoisomerizable chromophores for the size change of polymer solids was for the first time proposed by Merian (1966). He studied a nylon filament fabric, 6 cm wide and 30 cm long, dyed with 15 mg/g azo dye. After exposure to a xenon lamp at a distance of 30 cm, the dyed fabric had shrunk by 0.33 mm. Since this finding, many materials exhibiting photostimulated deformations have been reported. Polymer films mixed with low molecular weight photochromic compounds, such as nylon film–carotene, nylon film–cyanostilbene and polystyrene–spirobenzopyran, have been shown to undergo photostimulated reversible size changes (Blair and Law, 1980; Blair and Pogue 1982).

A covalently bound photochromic chromophore is expected to give a more direct effect on the deformation of polymers. The systems studied were polyimide with backbone azobenzene groups (Agolini and Gay, 1970), polyamide with backbone stilbene groups (Osada et al., 1981), polyquinoline with backbone stilbene groups (Zimmerman and Stille, 1985), poly-tetrahydrofuran with backbone viologen groups (Kohjiya et al., 1985), poly (ethyl acrylate) with spirobenzopyran or azobenzene groups as cross-linking agents (Smets and Evans, 1973; Smets et al., 1978; Eisenbach, 1980) and poly(n-butyl acrylate) with pendant azobenzene groups (Matějka et al., 1981).

The most pronounced photocontraction effect was observed for poly (ethyl acrylate) cross-linked with spirobenzopyran groups (**18**). This is shown in Figure 5.15.

Initially, it was assumed that the contraction was due to an entropy increase of the polymer chain associated with a higher flexibility of the open-ring merocyanine form compared with the parent ring-closed spiropyran. However, careful re-examination of the action spectrum of the contraction, i.e. the dependence of film shrinkage on the wavelength of irradiation, led to the unexpected finding that the contraction spectrum closely fitted with the absorption spectrum of open-form merocyanine and

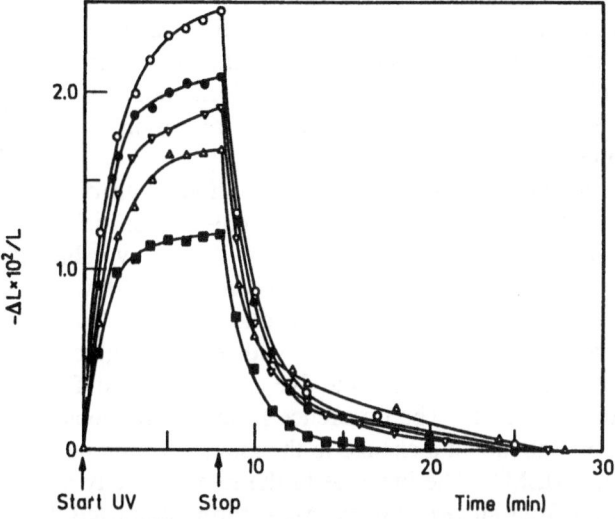

Figure 5.15 Photomechanical behaviour of ethyl acrylate cross-linked with spirobenzopyran groups. Influence of temperature and stress time in minutes.

Temperature	l (mm)	Load (g)
15	44.5	21.9
29.6	44	21.9
45	42	21.9
15	22	35.7
15	32.5	59.4

that UV light causing the isomerization from the spirobenzopyran to the merocyanine form was very inefficient for film contraction. Thus, the film shrinkage appeared to be induced not by the photochemical isomerization of the spirobenzopyran to the merocyanine form, but by the increase in local temperature arising from a non-radiative transition of the photoexcited merocyanine.

In order to minimize the local heating effect, Van der Veen and Prins (1971) proposed to experiment with solvent-swollen gels, in which rapid thermal conduction is supposed to suppress this unfavorable effect. The gel system used by Prins *et al.* was a mixture consisting of low molecular weight chrysophenin G (**1**) and a water-swollen gel of poly(2-hydroxyethyl methacrylate) (PHEMA) cross-linked with ethylene glycol dimethacrylate [1.1% (w/w)]. Upon irradiation, the dye changed the configuration from all *trans* to the *c–t–c* form. The difference in the hydrophobic property of the two forms brought about a change in the intermolecular interaction between the dye and PHEMA. The increase in the hydrophilicity of the dye by the isomerization from all *trans* to the *c–t–c* form contracted the polymer gel, because this action liberated the hydrophilic dyes from the polymer chain to the surrounding solution.

A similar photoeffect on the swelling of cross-linked poly(methacrylic acid) (PMA) was observed in the presence of 4-phenyl-azophenyl trimethylammonium ions (Chuang *et al.*, 1973).

Even when a solvent-swollen gel is used, there remains a question as to the relative contribution of the local heating and the real photochemical reaction to the observed photoshrinking. Matějka *et al.* (1979) and Matějka and Dušek (1981) carefully examined the contribution of the former to the photostimulated shrinkage of a maleic anhydride–styrene copolymer with covalently bound pendant azobenzene groups (MAH–STY–AAB) swollen in diethylphthalate. They measured the photogenerated force as well as the temperature of the gel by inserting a thermocouple into the gel. The copolymer was irradiated at an elongation of 1.25%. The irradiation caused a reversible increase in the force of 1% (Figure 5.16). As seen in the figure, the force change rate is much faster than the isomerization reaction, and the response correlates well with the change in the temperature of the gel. The latter fact suggests that the decisive role played in the contraction process is the local heating due to light absorption, and not apparently the photoisomerization of the photochromic chromophores. When intensely photoirradiated, the sample was heated by the absorption of radiation even under careful thermostatic control. The rise in temperature inside the gel was determined to be 1.2 K. With an interference filter ($\lambda = 370$ nm) it was possible to minimize the heating effect. The slow increase of the force as seen in Figure 5.16(b) may be ascribable to the photochemical *trans–cis* isomerization, though the effect is less than 1%.

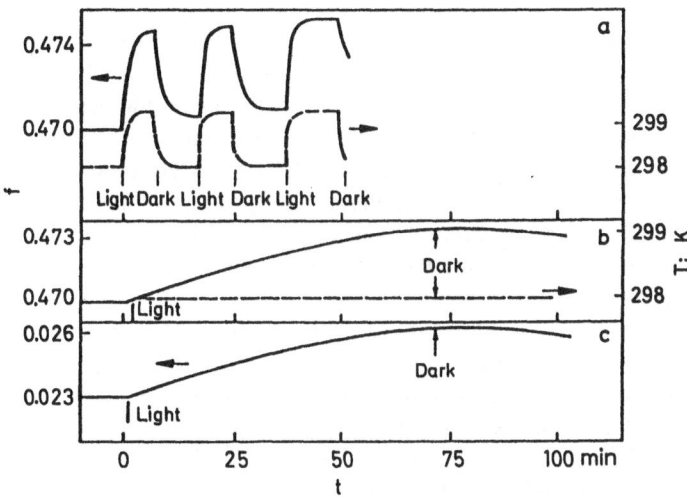

Figure 5.16 Effects of radiation on retractive force f (in newtons) at constant length of a sample of poly(MAH–STY–AAB) swollen in diethyl phthalate and on temperature T_i inside the sample. —, f; – – –, T_i. (a) $l = 1.25$ without interference filter, (b) $l = 1.25$ with interference filter, (c) $l = 1.05$ without interference filter.

The large heat effect observed even in the contraction of the solvent-swollen gel strongly suggests that many previous studies reporting observed photostimulated contractions should be re-examined to check and evaluate the real photochemical effect. The following criteria may be useful to judge whether the effect is due to photochemistry or photoheating.

(1) When the recovery rate of deformation in the dark after light is switched off is faster than the rate of thermal isomerization of the chromophores in the film, the contraction is due to photoheating. The recovery rate should always be slower than the rate of isomerization even when the recovery is induced by photoirradiation, if the process is associated with the photochemical reaction of the chromophores.

(2) At elongations l smaller than the inversion elongation l_{inv}, the temperature increase causes the force to decrease, while at l larger than l_{inv}, an increase in the modulus with temperature causes the contraction force to increase. Therefore, photoheating should give rise to reverse effects depending on l. On the other hand, the photochemical effect is independent of l. If the photostimulated contraction depends on l, the contraction is due to photoheating.

Although many systems showing photostimulated deformations have been reported, the deformations were limited to less than 10%. Small deformations make it difficult to judge whether the effect is due to photochemistry or photoheating. If the deformation is larger than 20%, it may safely be said that it is due to the photochemical effect. One photodeformable material of interest is poly-4-(N,N-dimethylamino)-N-r-D-glutamanilide, which displayed a dilation amounting to 35% in N,N-dimethylformamide when exposed to light in the presence of CBr_4 (Aviram, 1978). This large deformation was due to the ionization of N,N-dimethylanilide groups. Although the system was irreversible and no attempt was made to make it reversible, such a pronounced effect is informative for the design of reversible photodeformable polymers.

It may be inferred from the studies on the conformation change in solution that the electrostatic repulsion between photogenerated charges is more effective for conformation changes than the *trans–cis* isomerization of unsaturated linkages. On due consideration, the author decided to take advantage of electrostatic forces to obtain gels exhibiting large reversible deformations (Irie and Kungwatchakun, 1984, 1986).

Acrylamide gels containing a small amount of triphenylmethane leucohydroxide or leucocyanide groups (19) were prepared by free radical copolymerization of di(N,N-dimethylaniline)-4-vinylphenylmethane leucohydroxide (12, X=OH, R=CH=CH₂) or leucocyanide (12, X=CN, R=CH=CH₂) in dimethylsulfoxide in the presence of N,N-methylenebisacrylamide. The gels were swollen to equilibrium by allowing them to stand in water overnight. Then the changes in their weight and dimensions induced by UV light were measured.

A disk-shaped gel (10 mm in diameter and 2 mm thick) having 3.7% (mol/mol) triphenylmethane leucohydroxide residues showed photostimulated reversible dilation in water. Figure 5.17 shows that on UV irradiation ($\lambda > 270$ nm), the gel swells and its weight increases by as much as three times in 1 h. The dilated gel deswells in the dark to the initial weight in 20 h. The cycles of dilation and contraction of the gel were repeated several times. The gel having leucohydroxide residues swelled even in the dark when the aqueous solution became acidic owing to the chemical ionization of the residues.

In order to make the gel insensitive to pH changes, the hydroxide residues were replaced by cyanide groups. The weight of the leucocyanide gel remained constant in the range of pH 4–9. On UV irradiation, the gel weight increased as much as 18 times. In the dark, the gel contracted again slowly to the initial weight.

The triphenylmethyl cation is well known to have a very strong absorption at 622 nm. On UV irradiation, the color of the gel changes from pale-green to deep-green in less than 3 min and then remains almost constant. In the dark, the color returns to the initial pale-green in several hours. The size of the gel, on the other hand, increases slowly and reaches the saturated value in about 2 h. The slow response of the size change in comparison with the

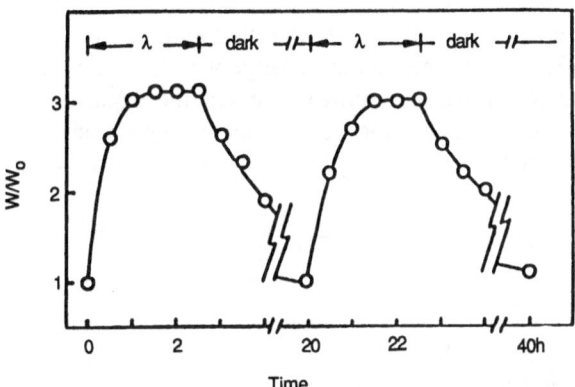

Figure 5.17 Photostimulated dilation and contraction of polyacrylamide gel having pendant triphenylmethane leucohydroxide groups [3.7% (mol/mol)] with light of wavelength longer than 270 nm at 25°C. W_0 is the weight before photoirradiation.

color change indicates that the gel dilation is due not to the thermal effect but to the photochemical ionization. The formation of charges, fixed cations and free anions, generates an osmotic pressure difference between the gel and the outer solution, and this osmotic effect is considered to be responsible for gel expansion. It is worthwhile to note that the gel expansion is suppressed by the addition of salts, such as sodium chloride and potassium bromide. No photostimulated dilation was observed in the presence of 10^{-2} M sodium chloride or potassium bromide for the gels having leucohydroxide or leucocyanide groups.

The acrylamide gel described above is the first example showing a reversible deformation of more than 100%. The effect is purely photochemical and reversible. However, this gel has a serious disadvantage in that the response time is slow. There are two approaches to resolve the deficiency. One approach is to reduce the gel size (Irie *et al.*, in press), while another is to employ the help of the electric field to increase the gel deformation rate (Irie, 1986). The following discussion shows the results of the two approaches.

A rod-shaped acrylamide gel containing triphenylmethane leucocyanide groups (initial radius, R_0, 11, 37, 90 and 180 µm; ratio of the length to the diameter, 3.6) was prepared in a capillary tube. The change in the diameter of the gels was measured as a function of irradiation time under a microscope equipped with a charged coupled device camera and a television monitor. Assuming that the volume change is isotropic in all directions of the gel, the volume ratio, V/V_0, where V_0 and V are the volume of the gel at $t = 0$ and $t = t$ respectively, was measured. The results are summarized in Figure 5.18. The photoresponse time of V/V_0 is strongly dependent on the initial radius of the gel, while the equilibrium V/V_0 at infinite irradiation time ($t = \infty$) is almost independent of the initial radius. It is worth noting that the smaller the size of the gel, the faster the photoinduced volume change. For the gel with R_0 of 11 µm the volume change was almost complete in 1 min, while that with R_0 of 180 µm showed the volume expansion over 30 min. The response time is much improved by decreasing the size of the gel.

The photoinduced volume change is analyzed by the following equation:

$$R_e - R(t) = (R_e - R_o)(6/\pi^2) \sum_{n=1}^{\infty} n^{-2} \exp(-n^2 t/\tau)$$

$$\tau = R_e^2/\pi^2 D$$

R_e and $R(t)$ are the radii of a gel at $t = \infty$ and $t = t$ respectively. The response time, τ, is related to the diffusion coefficient of the gel network, D. An equation has been proposed by Tanaka and Fillmore (1979) to analyze the swelling the process of a spherical acrylamide gel in water. When $t/\tau > 0.25$, the above equation is simplified as follows:

$$[R_e - R(t)]/(R_e - R_0) = A \exp(-t/\tau)$$

The author fitted log $[R_e - R(t)]$ *vs* t for the data shown in Figure 5.18 and obtained τ from the slopes. The results are summarized in Figure 5.19. This plot clearly demonstrates that τ is proportional to the square of R_e, indicating that τ is determined by the diffusion of the gel network ($D = 1.2 \times 10^{-7}$ cm^2/s).

An electric field is an effective stimulus for increasing the gel deformation rate. A rod-shape gel (25 mm in length and 1 mm in section radius) was placed between two parallel platinum electrodes in a small water pool (Teflon, 36 × 19 × 15 mm). Contact of the gel with the electrodes was avoided. Figure 5.20 shows the photostimulated bending motion of the acrylamide gel having 3.1% (mol/mol) triphenylmethane leucocyanide

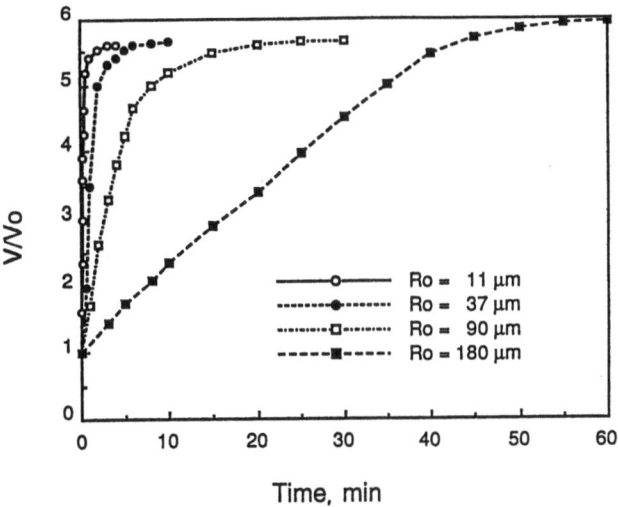

Figure 5.18 Size effect on the photoinduced volume change of the rod-shaped gel in water.

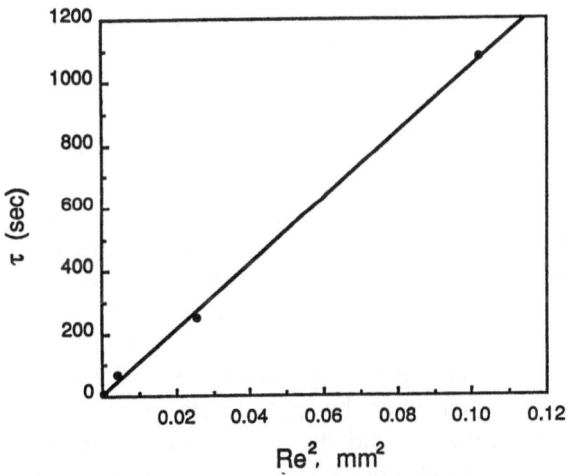

Figure 5.19 A relation between the response time t and R_e^2. (See text for details.)

groups in water in an electric field (Irie *et al.*, 1986). In the dark, the gel did not change shape in an electric field (10 V/cm). Upon UV irradiation, the gel quickly bent in 1 min. The gel end moved in the direction of a positive electrode. During the bending motion, the center of gravity of the gel remained at the initial position. Translational movement of the entire gel to the negative electrode was not observed. By changing the polarity of the electric field the gel again became straight and then bent to another direction.

(a)

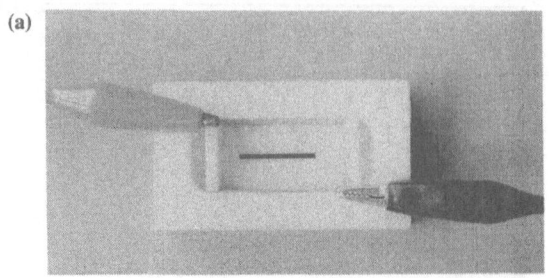

(b)

Figure 5.20 Photostimulated bending motion of a rod-shaped acrylamide gel(25 mm in length and 1 mm in radius) containing 3.1% (mol/mol) triphenylmethane leucocyanide groups in an electric field (10 V/cm). (a) Before photoirradiation; (b) under UV irradiation.

The response time of the gel shape change was around 2 min. After switching off the light, the gel slowly returned to the initial straight shape in the electric field.

In order to determine quantitatively the response time of the motion, one end of the rod-shaped gel was fixed to the wall and the distance that the free end moved from the initial position was measured as a function of irradiation time. Figure 5.21 shows the result. The free end moved toward the positive electrode with an initial speed of 0.40 mm/s. When the polarity of the electric field was changed, the gel end moved in the opposite direction.

The bending rate depended on the applied field and the salt concentration in water. Upon increasing the field, the end-moving rate increased in proportion to the square root of the applied field. The rate in the solution containing 2×10^{-4} mol/l magnesium chloride was 10 times faster than the rate in the absence of magnesium chloride. Under UV irradiation in the presence of salt the gel vibrated in response to the alternating electric field (+ 8.5 V/cm, 0.5 Hz) as shown in Figure 5.22.

The bending behaviour of the gel suggests inhomogeneous expansion of the gel in the electric field. The negative electrode side of the gel expands more than the other side. Since the electric field is applied perpendicular to the gel axis, mobile negative ions are attracted to the positive electrode side in the gel. Consequently, excess positive charges are left on the other side.

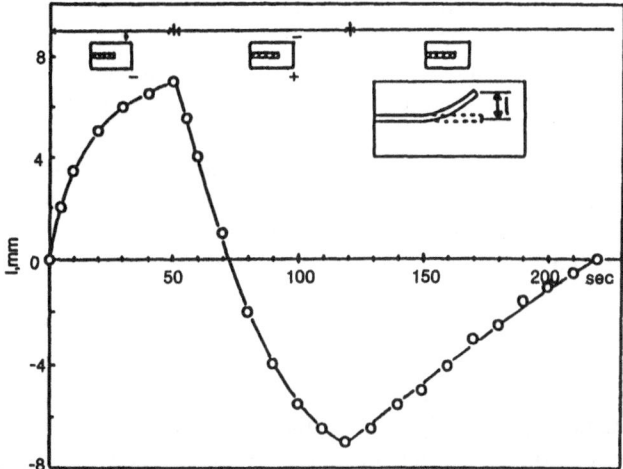

Figure 5.21 Photostimulated bending of a rod-shaped acrylamide gel (26 mm in length and 1 mm in section radius) having 3.1% (mol/mol) triphenylmethane leucocyanide groups in an electric field(10 V/cm) in water. The electric field was removed after 120 s.

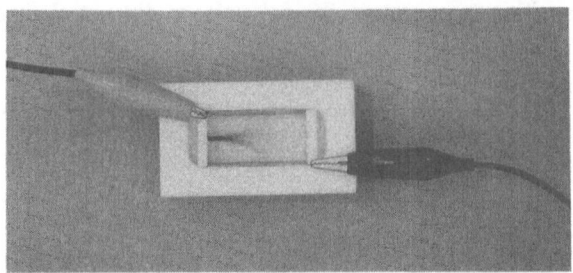

Figure 5.22 Photostimulated vibrational motion of a rod-shaped acrylamide gel having 3.1% (mol/mol) triphenylmethane leucocyanide groups under an alternating electric field (+ 0.8 V/cm, 0.5 Hz) in water in the presence of 4×10^{-4} M sodium chloride.

The large mobile ion concentration gradient on the gel surface is responsible for the gel bending. The thinness of the active layer, less than 10 μm as estimated from the behaviour of tiny rod-shaped acryamide gels described above, is considered to play an important role in increasing the response time.

5.5 Photostimulated volume phase transition of gels — discontinuous change

By introducing the concept of photostimulated phase transition as shown in Figure 5.7 to the gel system, it becomes possible to construct a sensitive photodeformable gel. Poly(*N*-isopropylacrylamide) gels having triphenyl-methane leucocyanide groups were synthesized in attempts to make a gel which shows a photostimulated volume phase transition (Manada *et al.*, 1990). The characteristic feature of a poly(*N*-isopropylacrylamide) gel with ionizable groups is that it undergoes a phase transition in pure water as the temperature is varied (Hirokawa and Tanaka, 1984; Hirotsu *et al.*, 1987). The gels having ionizable groups (sodium acrylate) ranging from 0 to 0.3% (mol/mol) undergo continuous transitions, whereas the gels with more than 0.6% (mol/mol) have discontinuous volume phase transition. This result suggests the possibility of photostimulated phase transition by introducing photoionizable chromophores into poly(*N*-isopropylacrylamide) gels.

Figure 5.23 shows equilibrium volumes of the gel measured as a function of temperature, with and without UV irradiation which induces the photodissociation of the leucocyanide groups. When the gel was not irradiated with UV light, it underwent a sharp, but continuous, volume change at 30.0°C. Upon irradiation the gel showed a discontinuous volume transition. The temperature was raised gradually from 25.0°C. At 32.6°C, the volume of the gel suddenly, decreased by approximately 10 times. Above the transition temperature the gel volume did not change markedly.

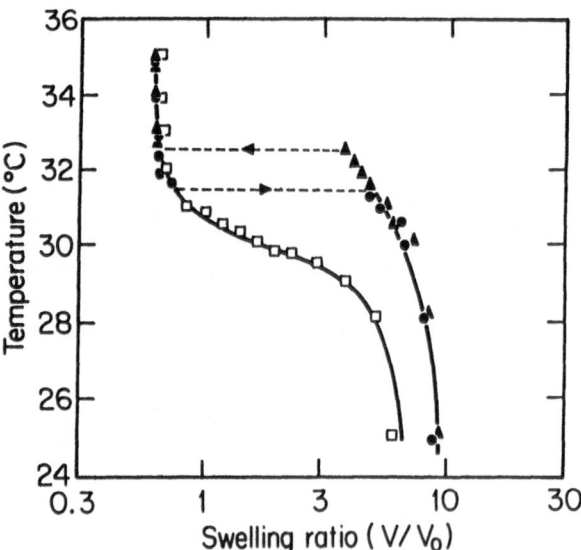

Figure 5.23 Degree of swelling of poly(*N*-isopropylacrylamide) gel having leucocyanide pendant groups in water as a function of temperature. Open squares show the swelling of the gel without UV irradiation, on both raising and lowering the temperature. Filled triangles denote the swelling curve on UV irradiation when raising the temperature. Filled circles are for the curve on UV irradiation when lowering the temperature.

When the temperature was lowered, starting from 35.0°C, the gel swelled discontinuously to 10 times its original size at approximately 31.5°C. The presence of the hysteresis confirms that this is the phase transition. When the temperature was fixed at 32.0°C, the gel underwent a discontinuous swelling–shrinking switching upon exposure to and removal of UV light.

In the present experiment, the color of the gel changed from pale-green to dark-green on UV irradiation. Therefore, when the gel was not irradiated, it was a non-ionic gel, but it became an ionic gel upon UV irradiation. As described above, an osmotic pressure induced by ionization of a gel brings the gel state below the coexistence curve and induces a discontinuous volume phase transition.

References

Angolini, F. and Gay, F.P. (1970) *Macromolecules* **3**, 349.
Aviram, A. (1978) *Macromolecules* **11**, 1275.
Blair, H.S and Law, T.K. (1980) *Polymer* **21**, 1475.
Blair, H.S. and Pogue H.I (1982) *Polymer* **23**, 779.
Blair, H.S. Pogue, H.I. and Riordan, J.E. (1980) *Polymer* **21**, 1195.
Chuang, J.C., de Sorgo, M. and Prins, W. (1973) *J. Mechanochem. Cell Motil.* **2**, 105.
Ciardelli, F., Carlini, C. Salaro, R. Altmore, A., Pieroni, O., Houben, J.L. and Fiss, A. (1984) *Pure Appl. Chem.* **56**, 329.
Eisenbach, C.D. (1980) *Polymer* **21**, 1175.
Fox, T.G. and Flory, P.J. (1951) *J.Am.Chem.Soc.* **73**, 1909, 1915.
Goldburt, E., Shvartsman, F., Fishman, S. and Krongauz, V. (1984) *Macromolecules* **17**, 1225.
Hampson, G.C. and Robertson, J.M. (1941) *J. Chem. Soc.*, 409.
Hartley, G.S. (1938) *J. Chem. Soc.*, 633.
Hirokawa, Y. and Tanaka, T. 1984 *J. Chem. Phys.* **81**, 6379.
Hirotsu, S., Hirokawa, Y. and Tanaka, T. (1987) *J. Chem. Phys.* **87**, 1392.
Irie, M. (1986) *Macromolecules* **19**, 2890.
Irie, M. (1990) *Pure Appl. Chem.* **62**, 1495.
Irie, M. and Hayashi, K. (1979) *J. Macromol. Sci. Chem* **A13**, 511.
Irie, M. and Hosoda, M. (1985) *Makromol. Chem. Rapid Commun.* **6**, 533.
Irie, M. and Kungwatchakun, D. (1984) *Mokromol. Chem. Rapid Commun.* **5**, 829.
Irie, M and Kungwatchakun, D. (1986) *Macromolecules* **19**, 2476.
Irie, M. and Schnabel, W. (1985) *Macromolecules*, **18**, 394.
Irie, M. and Suzuki, T. (1987) *Makromol. Chem. Rapid Commun.* **8**, 607.
Irie, M. and Tanaka, H. (1983) *Macromolecules* **16**, 210.
Irie, M., Hirano, K. Hashimoto, S. and Hayashi, K. (1981a) *Macromolecules* **14**, 262.
Irie, M., Menju, A. and Hayashi, K. (1981b) *Macromolecules* **12**, 1176.
Irie, M., Iwayanagi, T. and Taniguchi, Y. (1985) *Macromolecules* **18**, 2418.
Ishihara, K., Hamada, N., Kato, S. and Shinohara, I. (1984) *J. Polym. Sci. Chem. Ed.* **22**, 121.
Kumar, G.S., DePra, P. and Neckers, D.C. (1984a) *Macromolecules* **17**, 1912.
Kumar, G.S., DePra, P., Zhang, K. and Neckers, D.C. (1984b) *Macromolecules* **17**, 2463.
Kumar, G.S., Savariar, C., Sattran, M. and Neckers, D.C. (1985) *Macromolecules* **18**, 1525.
Kohjiya, S., Hashimoto, T., Yamashita, S. and Irie, M. (1985) *Chem. Lett.*, 1479.
Kungwatchakun, D. and Irie, M. (1988) *Makromol. Chem. Rapid Commun.* **9**, 243.
Lovrien, R. (1967) *Proc. Nat. Acad. Sci. USA* **57**, 236.
Mamada, A., Tanaka, T., Kungwatchakun, D. and Irie, M. (1990) *Macromolecules* **23**, 1517.
Matějika, L., Dŭsek, K. and Ilavský, M. (1979) *Polym. Bull.* **1**, 659.

Matějika, L., Ilavský, M., Dušek, K. and Wichterle, O. (1981) *Polymer* 22, 1511.
Matějika, L. and Dušek, K. (1981) *Makromol. Chem.* 182, 3223.
Menju, A., Hayashi, K. and Irie, M. (1981) *Macromolecules* 14, 755.
Merian, E. (1966) *Text. Res. J.* 36, 612.
Negishi, N., Takahashi, M., Iwazawa, A., Matsuyama, K. and Shinohara, I. (1977) *Nippon Kagaku Kaishi*, 1035.
Negishi, N., Isihihara, K. and Shinohara, I. (1982) *J. Polym. Sci., Chem Ed.* 20, 1907.
Osada, Y., Katsumura, K. and Inoue, K. (1981) *Makromol. Chem. Rapid Commun.* 2, 47.
Smets, G. and Evans, G. (1973) *Pure Appl. Chem. Macromol. Chem.* 8, 357.
Smets, G., Breaken, J. and Irie, M. (1978) *Pure Appl. Chem.* 50, 845.
Tanaka, T and Fillmore, D.J. (1979) *J. Chem. Phys.* 70, 1214.
Ueno, A. and Osa, T. (1980) *Yuki Gosei Kagaku* 38, 267.
Van der Veen, G. and Prins, W. (1971) *Nature Phys Sci.* 230, 70.
Van der Veen, G. and Prins, W. (1974) *Photochem. Photobiol.* 19, 191.
Zimmerman, G., Chow, L. and Paik, U. (1958) *J. Am. Chem. Soc.* 80, 3528.
Zimmerman, E.K. and Stille, J.K. (1985) *Macromolecules* 14, 1246.

6 High and low molecular weight photochromic viologen-based systems

H. KAMOGAWA

6.1 Introduction

Viologens are widely known as a promising class of organic electrochromic substances (Schoot *et al.*, 1973). However, few studies of their photo-excitation and subsequent reduction with appropriate reductants have been reported, except for those involving extremely high-speed techniques such as flash photolysis, presumably because of the highly oxidation-sensitive nature of the radical cation produced as a result of one-electron reduction (Johnson and Gutowsky, 1963; Takuma *et al.*, 1977; Kohjiya *et al.*, 1985).

This chapter relates to the reduction of viologens and the resulting visible photochromism principally in the solid state induced by light of the wavelengths and intensities encountered with the ordinary sunlight in our ambient environment. Use of matrix polymers or copolymers bearing viologen units is usually required to develop colours by light in the film state. However, it has also been found that some types of viologen exhibit photochromism in their crystalline state. The viologens under discussion in this chapter are denoted **I, II** and **III**.

The dicationic part of viologen undergoes reduction according to eqn (1):

$$\tag{1}$$

6.2 Synthesis of viologens bearing vinyl groups and their polymers

4,4′-Bipyridinium salts bearing a vinyl group, such as the monomers **V**, **IX**, **XIa–c**, and **XIII**, were synthesized from 4,4′-bipyridine (**I**) via its monopyridinium salt (**III**) according to eqns (2–5) (see Kamaogawa *et al.*, 1979). Thus, **III** was synthesized from **I** and an equivalent of propyl bromide

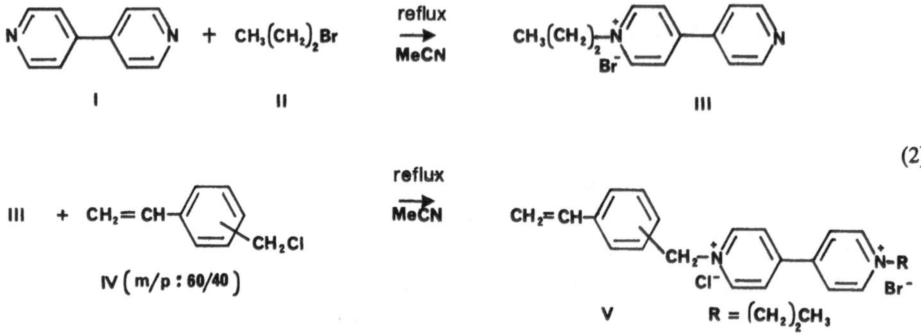

$$\text{I} \quad + \quad CH_3(CH_2)_2Br \quad \xrightarrow[\text{MeCN}]{\text{reflux}} \quad CH_3(CH_2)_2-N^+ \text{...} N \quad Br^-$$

$$\quad \text{I} \qquad\qquad \text{II} \qquad\qquad\qquad \text{III}$$

$$\tag{2}$$

$$\text{III} \quad + \quad CH_2=CH-\langle\text{aryl}\rangle-CH_2Cl \quad \xrightarrow[\text{MeCN}]{\text{reflux}} \quad CH_2=CH-\langle\text{aryl}\rangle-CH_2-N^+\cdots N^+-R$$

$$\text{IV } (m/p : 60/40) \qquad\qquad \text{V} \quad Cl^- \qquad Br^- \qquad R = (CH_2)_2CH_3$$

$$\text{III} \quad + \quad Br(CH_2)_2NH_2\cdot HBr \quad \xrightarrow[\text{MeCN}]{\text{reflux}}$$

$$HBr\cdot NH_2(CH_2)_2-N^+\cdots N^+-(CH_2)_2CH_3 \quad + \quad CH_2=CHCOCl \quad \xrightarrow{Et_3N}$$
$$\qquad Br^- \qquad\qquad Br^-$$

$$\text{VII} \qquad\qquad\qquad \text{VIII}$$

$$CH_2=CHCNH(CH_2)_2-N^+\cdots N^+-(CH_2)_2CH_3$$
$$\quad\;\; \overset{\parallel}{O} \qquad Cl^-\,(Br^-) \qquad Br^-\,(Cl^-)$$

$$\text{IX}$$

$$\tag{3}$$

$$\text{III} \quad + \quad CH_2=CRCO(CH_2)_nBr \quad \xrightarrow[\text{MeCN}]{\text{reflux}}$$
$$\qquad\qquad \overset{\parallel}{O}$$

X_a; R = CH_3 , n = 2
X_b; R = H   , n = 3
X_c; R = CH_3 , n = 3

$$CH_2=CRCO\,(CH_2)_n-N^+\cdots N^+-(CH_2)_2CH_3$$
$$\quad\;\; \overset{\parallel}{O} \qquad\qquad Br^- \qquad\qquad Br^-$$

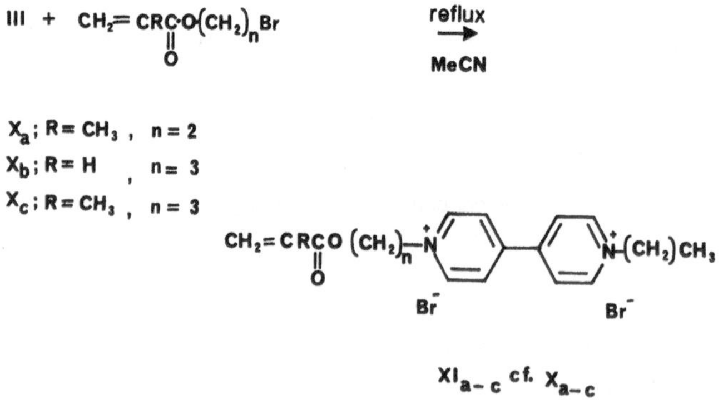

$$\tag{4}$$

$$XI_{a-c} \;\; \text{c.f.} \;\; X_{a-c}$$

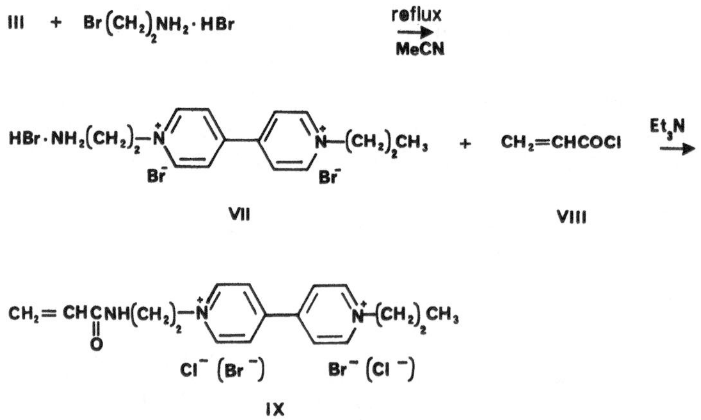

$$\text{III} + \text{ClCH}_2\text{C OCH=CH}_2 \xrightarrow[\text{MeCN}]{\text{reflux}} \text{CH}_2\text{=CHOCCH}_2\overset{+}{\text{N}}\!\!\underset{\text{Cl}^-}{\underbrace{\hspace{1cm}}}\!\!\overset{+}{\text{N}}(\text{CH}_2)\text{CH}_3 \quad (5)$$

XII   **XIII**

(II) in refluxing acetonitrile by a Menschutkin reaction. Styrenic (V), (meth)acrylic (IX, XIa–c) and unconjugated (XIII) vinyl double bonds were attached to III by quaternizing the remaining pyridine moiety. The low yield achieved (12%) for XIa was presumably due to possible dehydrobromination.

Recently, viologen (ViV^{2+}) in which a vinyl group is attached directly to one of the bipyridinium nitrogens was reported (Nambu et al., 1986). Viologens bearing a long spacer between the vinyl group and bipyridinium nitrogen have also been reported (Tundo et al., 1982).

$$\quad (6)$$

$$\text{V,V}^{2+} \qquad\qquad \begin{aligned} 1&: X = CH_3 \,;\; Y = CO_2CH_3 \\ 2&: X = H \;\;\,;\; Y = CN \end{aligned}$$

Table 6.1 (see Kamogawa et al., 1979) summarizes the polymerization behaviour of the viologen monomers V, IX, XIa–c and XIII when aqueous solutions (5 ml) of monomers (1 g in total) and ammonium persulphate (0.01 g as radical initiator) were allowed to stand at 80°C for 24 h under nitrogen. It can be seen from this table that homopolymerization of viologen monomers gives lower values (entries 1, 3, 7) of both conversion and intrinsic viscosity than those for their copolymers with acrylamide (entries 2, 4, 8), except for entries 9, 10 and 11. Conversion for the copolymerization of the monomer XIII is low (entry 10, 7%), presumably because of poor copolymerizability, analogous to vinyl acetate.

Table 6.1 Polymerization behaviour of viologen monomers.

No.	Monomer	Conversion (%)	[η] (dl/g)[a]	Elemental analysis[b] (%)	Remark
1	**V**	26	0.44	C, 61.85 (61.19); H, 5.78 (5.60); N, 5.97 (6.49)	Viologen homopolymer
2	**V-AM**[c]	43	0.86	C, 59.07 (59.71); H, 5.71 (5.81); N, 6.75 (8.36)	Viologen copolymer
3	**IX**	32	0.51	C, 47.51 (47.28); H, 5.08 (5.07); N, 9.02 (9.19)	Viologen homopolymer
4	**IX-AM**[c]	65	1.85	C, 54.80 (53.96); H, 5.45 (6.00); N, 11.07 (11.99)	Viologen copolymer
5	**XIa**	20	0.27	C, 50.11 (48.28); H, 5.13 (5.15); N, 5.33 (5.97)	Viologen homopolymer
6	**XIb**	31	0.38	C, 47.23 (46.97); H, 5.30 (5.26); N, 5.71 (6.07)	Viologen homopolymer
7	**XIc**	23	0.26	C, 49.82 (49.40); H, 5.15 (5.39); N, 5.38 (5.76)	Viologen homopolymer
8	**XIc-AM**[c]	35	1.19	C, 53.57 (50.13); H, 5.16 (6.70); N, 6.94 (6.92)	Viologen copolymer
9	**XIII**	23	—	N, 1.32 (7.02)	Substantially PVA
10	**XIII-AM**[c]	7	—	N, 10.43 (8.94)	Substantially PVA–AM copolymer
11	**XIII**	19[d]	—	N, 1.12	Viologen graft copolymer

[a] Intrinsic viscosity in aqueous solution at 25°C.
[b] Figures in parentheses indicate the values calculated for the monomers used.
[c] 1:1M ratio, AM = acrylamide.
[d] Graft copolymerization onto PVA.

Table 6.2 Redox properties of viologen polymers.

Compound	E_0(mV) First	E_0(mV) Second	Absorption peak[a] a by reduction (nm)	Absorption peak[b] by UV irradiation (nm)
Monomer **V**	194	−336	600	—
V homopolymer	324	−276	530	610
V–AM copolymer[c]	214	−266	530	610
Monomer **IX**	199	−340	600	—
IX homopolymer	282	−289	530	610
Monomer **XIc**	194	−366	600	—
XIc homopolymer	244	−276	530	610
XIc–AM copolymer[c]	284	−306	530	610
Monomer **XIII**	294	−336	600	—
XIII graft[d] copolymer	—	—	600	600

[a] With $Na_2S_2O_4$–NH_3 in aqueous solution.
[b] Mercury lamp (75 W) irradiation on film (10 min).
[c] Copolymers in Table 6.1.
[d] No. 11 graft copolymer on PVA in Table 6.1.

Table 6.2 (Kamogawa *et al.*, 1979) indicates standard reduction potentials (E_0) of both viologen monomers and polymers by means of reductive titration and visible absorption peaks of radical cations produced as a result of reduction. It is interesting to note that the absorption peaks for polymers developed in solutions with reductant are different from those in the film state, except for **XIII**, the graft copolymer, which, as will be described later, may be the result of aggregation of the radical cations produced in the former.

The results of copolymerization of monomer ViV^{2+} are also given in Table 6.3 (Nambu *et al.*, 1989).

Table 6.3 Radical copolymerization of viologen monomers with MMA and AN[a].

V^{2+} monomer (mol % in feed)	Comonomer	Copolymer (yield, %)	Elemental analysis of nitrogen (%)	Content of V^{2+b} [% (mol/mol)]
ViV^{2+} (10)	MMA	**1a** (28)	1.37	5.8
ViV^{2+} (15)	MMA	**1b** (20)	2.02	9.4
ViV^{2+} (20)	MMA	**1c** (13)	3.07	17.0
ViV^{2+} (25)	MMA	**1d** (11)	4.72	37.3
ViV^{2+} (5)	AN	**2a** (43)	19.92	5.7
ViV^{2+} (15)	AN	**2b** (29)	15.21	13.9
$ViBzV^{2+}$ (3)	MMA	**3a** (18)[c]	0.64	2.5
$ViBzV^{2+}$ (20)	MMA	**3b** (10)[c]	2.87	17.9

[a] Run as 30 M. 3 M solutions in γ-butyrolactone with 2% (mol/mol) of AIBN at 80°C for 24 h. Polymers precipitated with MeOH.
[b] Estimated by elemental analysis of nitrogen.
[c] Run as 3 M solutions at 60°C for 20 h.

6.3 Reversible colour development (photochromism) of viologens by photoreduction mechanism (in matrices)

As already shown in the last column of Table 6.2, the generation of radical cations from viologens by one-electron reduction can be effected by the application of light. It is known that, upon irradiation with sunlight, viologens in ethyl or isopropyl alcohol turn pale-blue in the absence of oxygen, indicating the production of radical cations by one-electron reduction of viologens with alcohol as reductant (Johnson and Gutowsky, 1963).

The reversible photoreduction and the corresponding intense colour development of viologens (photochromism) in thin polymer matrix films were first investigated by Kamogawa *et al.* (1980). Table 6.4 indicates the type of viologen prepared in the previously described manner (Menschtkin reaction). The matrix polymers employed are also shown in Table 6.5. Generally, a matrix polymer (0.5 g) was dissolved in water (5 ml). A viologen was then dissolved in the solution to give 0.099 mol/l concentration.

The resulting solution was spread over a glass plate, followed by drying overnight at room temperature. The plate deposited with polymer film *ca* 0.1 mm thick thus prepared was then stored in a desiccator with minimum 30% relative humidity (RH) at least overnight before use. When the plate was taken out of the desiccator it was immediately irradiated at a distance of either 5 or 15 cm from a 75-W high-pressure mercury lamp (Toshiba SHL-100 UV) and the resulting change of absorption spectrum (colour development) was recorded on a spectrophotometer.

Table 6.6 summarizes the results of the effect of matrix polymer on colour development. Thus, an absorption with λ_{max} at 610 nm (blue) appeared more or less rapidly upon UV irradiation and, when irradiation was stopped, the colour developed was faded out to the original almost colourless state, presumably because of air oxidation of the coloured species. This colour development by light and subsequent fading in the dark can be cycled many times, and this phenomenon is considered to be 'photochromism by redox

Table 6.4 Viologens synthesized and used.

Compound	R	R'	X⁻
1a	$n\text{-}C_3H_7$	$n\text{-}C_3H_7$	Br⁻
1b	$n\text{-}C_3H_7$	$n\text{-}C_3H_7$	Cl⁻
1c	$n\text{-}C_3H_7$	$n\text{-}C_3H_7$	I⁻
2	PhCH₂	$n\text{-}C_3H_7$	Br⁻
3a	PhCH₂	PhCH₂	Br⁻
3b	PhCH₂	PhCH₂	Cl⁻
3c	PhCH₂	PhCH₂	BF_4^-
4	CH₃OOCCH₂	CH₃OOCCH₂	Cl⁻

Table 6.5 Polymers used as matrix.

Polymer	Abbreviation	Average MW	Remark
Poly (acrylamide)	PAM	$[\eta] = 1.9$ dl/g[a]	Aqueous polymerization
Poly (*N,N*-dimethyl acrylamide)	PMAM	$[\eta] = 0.8$ dl/g[a]	Ethanolic polymerization
Poly (vinyl alcohol)	PVA	20 000	Iwai Chem. Co.
Poly (*N*-vinyl-2-pyrrolidone)	PVP	10 000	Tokyo Kasei Co.

[a] Intrinsic viscosity in water at 30°C.

mechanism', as will be confirmed later. As can be seen clearly in Table 6.6, poly(N-vinyl-2-pyrrolidone)(PVP) provides, among others, remarkably high photosensitivities for the viologens **1a** and **3b**.

Figure 6.1 depicts typical visible absorption spectra for this photochromic system and also shows the outstanding high photosensitivity for viologen **3a** embedded in PVP matrix film.

Since it has been confirmed that PVP provides the highest photosensitivity in this viologen photochromic system, the effect of the type of viologen anion on the photochromism was investigated in PVP matrices. The results thus obtained are shown in Figure 6.2. In this figure, the photosensitivity decreases in the order: Cl⁻ > Br⁻ > I⁻, the order being reversed in the case

Table 6.6 Effect of matrix on colour development.

| Viologen | Matrix | Irradiation distance (cm) | Absorbance^a at irradiation time | | | $t_{1/2}$^{b} |
			30s	60s	90s	
1a	PVA	5	0.12	0.24	0.24	60
	PAM	5	0.00	0.04	0.04	40
	PMAM	5	0.13	0.38	0.56	10
	PVP	5	0.49	1.18	1.47	10
3b	PVA	15	—	0.13	—	600
	PVP	15	—	0.90	—	70

^{a} Value at absorbance maximum (610 nm).
^{b} Half-recovery time (min).

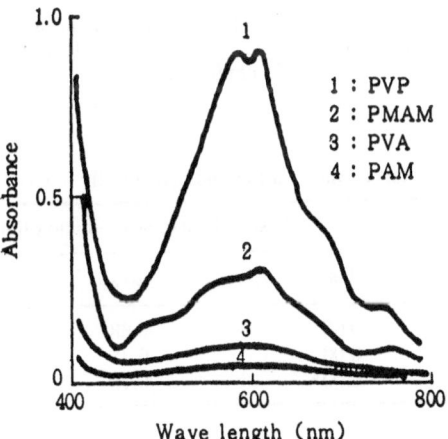

Figure 6.1 Effect of matrix on the colour development of the film containing viologen no. 3a (see Table 6.4) and exposed to UV light for 90 s at the distance of 15 cm from the lamp.

of nucleophilic attack in protic solvents such as methanol, but corresponding with that in polar aprotic solvents such as N,N-dimethylformamide (DMF). The same relationship is observed in Table 6.7, where the effect of BF_4^-, a stabilized anion, on the photosensitivity is shown.

Given that (1) PVP is hardly considered to become an electron donor for photoreduction and (2) the effect of the viologen counter anion on photosensitivity is significant and its order corresponds with that in polar aprotic

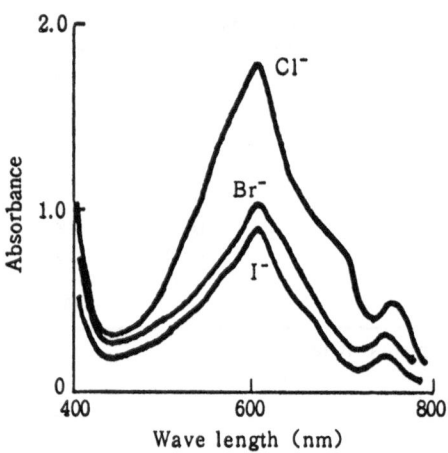

Figure 6.2 Effect of the counter anion in viologen on the colour development of PVP films bearing viologens 1a, 1b and 1c (see Table 6.4) and exposed to UV light for 60 s at the distance of 5 cm from the lamp.

Table 6.7 Effect of the viologen anion on the photochemically induced reduction[a].

	Absorbance[b] at irradiation time[c]		
Viologen[d]	0 s	20 s	60 s
3a	0.00	0.88	1.58
3c	0.00	0.35	0.88
1c	0.00	0.16	0.58

[a] Matrix PVP.
[b] Value at absorbance maximum (610 nm).
[c] 5 cm from UV lamp.
[d] Refer to Table 6.4.

solvents, the effect of matrix PVP on the photoreduction of viologen appears to be similar to that of N-methyl-2-pyrrolidone (MP), the corresponding polar aprotic solvent. For the 4,4'-viologen I, the effect of the R groups on the photosensitivity was also investigated.

The results thus obtained are given in Table 6.8. It is clear from the data in this table that viologens with less electron-donating N-substituents (R) such as benzyl (**3a**, Table 6.4) provide higher sensitivities than those bearing an electron-donating R group such as n-propyl (**1a**, Table 6.4), the order corresponding with that of the dark oxidation potential, as expected.

Table 6.8 Effect of N-substituent (R and R') on reduction rate[a].

	Absorbance[b] at irradiation time[c]			
Viologen[d]	20 s	40 s	60 s	$A_{20}/A_{60} \times 100$[e]
1a	0.40	1.18	1.58	25
2	0.71	—	1.50	47
3a	0.88	1.50	1.58	56
4	0.52[f]	0.63[f]	1.07[f]	49

[a] Matrix PVP.
[b] Value at absorbance maximum.
[c] 5 cm from UV lamp.
[d] Refer to Table 6.4.
[e] A, absorbance.
[f] At 605 nm (absorbance maximum).

Table 6.9 summarizes the effect of viologen type on photocolour development when embedded in PVP matrix film. It is apparent from this table that the extent of photoreduction as determined by the absorbance at 610 nm (the last column) parallels the values of the first dark redox potential; thus, for a given anion X, the less electron-donating R and R' afford larger values of the effect and, for given R and R' groups, the effect decreases for softer anions in the order: $Cl^- > Br^- > I^-$ and BF_4^-, thereby coinciding with the results already presented in Tables 6.7 and 6.8.

Table 6.9 Effect of viologen type on photocolour development (PVP matrix).

$$R-\overset{+}{N}\underset{X^-}{\diagup}\!\!\diagup\!\!\diagup\underset{X^-}{\overset{+}{N}}-R' \quad \text{Viologen}$$

No.	R	R′	X	First dark redox potential (mV)	Absorbance (610 nm)[a] immediately after irradiation
1	C_3H_7	C_3H_7	Cl	+29	1.25
2	C_3H_7	C_3H_7	Br	−6	1.18
3	C_3H_7	C_3H_7	I	−86	0.95
4	$PhCH_2$	$PhCH_2$	Cl	+89	> 2 (0.80)[b]
5	$PhCH_2$	$PhCH_2$	Br	+54	1.33
6	$PhCH_2$	$PhCH_2$	BF_4	−266	0.54
7	$PhCH_2$	C_3H_7	Br	+9	1.28
8	C_3H_7	$C_{14}H_{29}$	Br	+24	0.17
9	$CH_2=CHCH_2$	$CH_2=CHCH_2$	Br	−246	0.59
10	$MeOOCCH_2$	$MeOOCCH_2$	Cl	−40	1.17
11	$PhCH_2CO$	$PhCH_2CO$	Br	—	0.16
12	Me	Me	Cl	—	0.89

[a] Irradiation time 60 s at a distance of 5 cm from a 75-W mercury lamp. Absorbances before irradiation were all less than 0.10.
[b] Irradiation distances 15 cm.

6.4 Generation of viologen radical cations in solution

As described in section 6.3, the function of PVP in redox photochromism is to simulate the effect observed in polar aprotic solvents. The effect of various solvents (aprotic and protic) on the colour development of viologens was investigated and the results are shown in Tables 6.10 and 6.11 (Kamogawa and Amemiya, 1984; Kamagowa et al., 1984). Visible absorption spectra of the colours developed by light for viologens **1** and **4** (Table 6.9) under various conditions are also shown in Figure 6.3.

Intense colour development with an absorption peak around 610 nm for the viologen irradiated by light in the above-mentioned film state is attributable to the radical cation produced by the one-electron reduction of the viologen dication, as indicated in eqn (1).

It can be seen from Figure 6.3 that the shapes and the λ_{max} (610 nm) of the visible absorption curves 2 and 3 for solutions are quite similar to those of curve 1 for PVP matrix film, thereby additionally implying the generation of the radical cation in the latter case.

Table 6.10 Effect of solvent on photocolour development of viologens.

No.	Viologen[b]	Polar aprotic solvent[c]	Solubility of viologen	Absorbance (610 nm)[a]	
				A_0	A_t
1.1	4	MP	Very poor	0.29	0.07
1.2	4	MP/H$_2$O (9:1)[d]	Significant	0.12	0.84
1.3	4	MP/H$_2$O (1:1)[d]	Good	0.00	0.02
1.4	4	H$_2$O	Excellent	0.00	0.00
2.1	5	MP	Very poor	1.64	1.95
2.2	5	MP/H$_2$O (9:1)[d]	Significant	0.15	1.77
2.3	5	MP/H$_2$O (1:1)[d]	Good	0.00	0.42
3.1	4	HMPA	Very poor	0.02	0.20
3.2	4	HMPA/H$_2$O (9:1)[d]	Significant	0.03	0.24
4.1	5	HMPA	Very poor	0.26	0.19
4.2	5	HMPA/H$_2$O (9:1)[d]	Significant	0.15	1.10
5.1	4	DMF	Significant	0.02	0.12
5.2	4	DMF/H$_2$O (9:1)[d]	Good	0.02	0.18
6.1	5	DMF	Significant	0.01	0.27
6.2	5	DMF/H$_2$O (9:1)[d]	Good	0.01	0.36
7.1	4	DMF	Significant	0.09	0.09
7.2	4	DMA/H$_2$O (9:1)[d]	Good	0.01	0.05
8	5	DMA	Significant	0.09	1.25
9.1	4	DMSO	Good	0.00	0.02
9.2	4	DMSO/H$_2$O (9:1)[d]	Excellent	0.00	0.00
10	5	DMSO/H$_2$O (9:1)[d]	Excellent	0.00	0.01

[a] A_0, before; A_t, after 1 min irradiation.
[b] Numbers as given in Table 6.9.
[c] MP, N-methyl-2-pyrrolidone; HMPA, hexamethylphosphoric triamide; DMF, N,N-dimethylformamide; DMA, N,N-dimethylacetamide; DMSO, dimethylsulphoxide.
[d] Volume ratio.

Table 6.11 Effect of solvent on photocolour development of viologens.

No.	Viologen[b]	Protic solvent	Solubility of viologen	Absorbance (610 nm)[a]	
				A_0	A_t
11	5	MeOH	Excellent	0.00	0.30
12.1	4	i-PrOH	Good	0.00	0.52
12.2	4	i-PrOH/H$_2$O (9:1)[c]	Excellent	0.00	0.02
12.2					
13	5	i-PrOH	Poor	0.14	0.89
14.1	4	HCONH$_2$	Good	0.00	0.01
14.2	4	HCONH$_2$/H$_2$O (9:1)[c]	Excellent	0.00	0.00
14.2					
15	5	HCONH$_2$	Good	0.00	0.19
16	4	HCONHMe	Good	0.01	0.01
17	5	HCONHMe	Good	0.00	0.30

[a] A_0, before irradiation; A_t, after 1 min irradiation.
[b] Numbers as given in Table 6.9.
[c] Volume ratio.

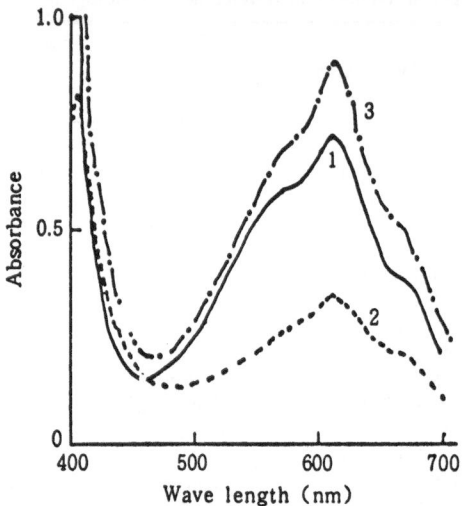

Figure 6.3 Visible absorption spectra for the viologen radical cations produced in various media. Curve 1, viologen no. 1 (see Table 6.9) in poly(N-vinyl-2-pyrrolodone) (PVP) matrix film, immediately after 20 s irradiation at a distance of 5 cm from a 75-W high-pressure mercury lamp; curve 2: viologen no. 4 (see Table 6.9) in N-methyl-2-pyrrolidone (MP) in the dark under nitrogen; curve 3, viologen no. 4 (see Table 6.9) in MP–water (9:1, v/v) under nitrogen, immediately after 60 s irradiation at a distance of 10 cm from the lamp.

From Table 6.10 it can be seen that from the various polar aprotic solvents (whether or not water is added) the mixed solvent MP — water (9:1, v/v) generates the highest value of $(A_t - A_0)$ values (see entries 1.2 and 2.2). The fact that anhydrous MP brings about spontaneous colour development and affords poor $(A_t - A_o)$ values appears to indicate that viologens in this solvent are too highly activated, so that the reduction in eqn (1) may proceed beyond the radical cation to the weakly coloured two-electron reduction product.

Since films which are cold cast from aqueous PVP contain about 10% water, the behaviour of viologens **4** and **5** (see Table 6.9) in the model solvent MP–water (9:1) may simulate the role of the PVP matrix in this type of photochromism.

It may be further noted from Table 6.10 that the highest photosensitivity is obtained with MP, the low molecular weight model for the polymer PVP, in which the solubilities of the viologens are very poor, thereby suggesting the presence of viologen ion pairs (Johns and Matthew, 1986). Polar aprotic solvents offering good solubilities for viologens such as dimethylformamide (DMF) and dimethylsulphoxide (DMSO) were found to provide poor results, which might also indicate that ion pairs play an important role in this type of photochromism.

The effects of the various protic solvents listed in Table 6.11 are complicated because of the possible reduction of viologens with solvent as electron donor. However, under the same conditions as employed with the aprotic solvents in Table 6.10, the photosensitivities are generally lower. The fact that *i*-propanol (entry 13), in which viologen **5** (see Table 6.9) is hardly soluble, gives a significant result might suggest also the presence of viologen ion pairs.

6.5 Photochromism of viologen crystals

Judging from the photochromic behaviour of viologens embedded in the PVP matrix film as well as in the solvent MP, the photoelectron transfer to viologen dications from the counterions is considered to contribute, at least partly, to the generation of the highly coloured viologen radical cations. To demonstrate this conjecture further, the redox photochromism of some viologen crystals without any added reductant has been described by Kamogawa *et al.* (1985) and Kamogawa and Suzuki (1987). Thus some of the viologens bearing sulphonate ions as the counterion exhibit redox photochromism in the crystalline state under UV light, irrespective of the absence or presence of air oxygen. Table 6.12 summarizes the results of the colour development obtained by subjecting various viologen crystals to 90 s irradiation at the distance of 10 cm from a 75-W high-pressure mercury lamp.

Figure 6.4 gives the visible reflection spectra indicating photocolour development for typical viologen disulphonate crystals **1a** and **6** (see Table 6.12).

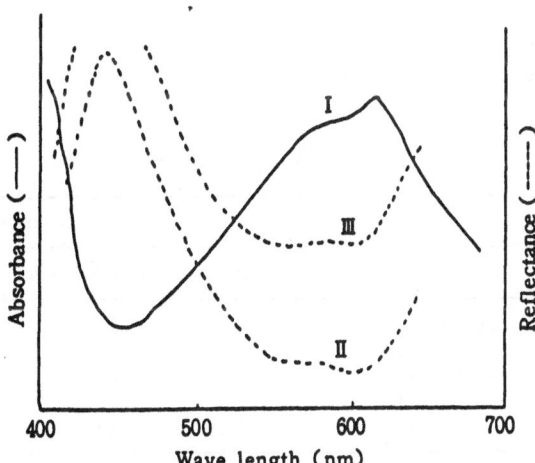

Figure 6.4 Reflection curves of the colours developed on viologen crystals with irradiation of near-UV light (> 350 nm). Curve I, absorbance for viologen no. 6 in Table 6.12; curves II and III, reflectances for viologen no. 1a in Table 6.12 under vacuum, immediately after 90 s irradiation and 10 min after respectively (colour sensor).

Table 6.12 Colour development of viologen crystals with illumination.

$$R-\overset{+}{N}\diagbndA\diagbnd\overset{+}{N}-R \quad 2X^-$$

Viologen no.	A	R	X	Colour development[a] In air[b]	Under vacuum[b]
1	None	PhCH$_2$	Br	None	None
1a			PTS[c]	Blue (610 nm)	Blue (610 nm)
1b			Benzene sulphonate	None	None
1c			MeSO$_3$	None	None
1d			1-Naphthalene-sulphonate	None	None
2	None	p-TolylCH$_2$	Br	None	None
2a			PTS	None	None
3	None	m-TolylCH$_2$	Br	None	None
3a			PTS	None	None
4	None	C$_3$H$_7$	Br	None	None
4a			PTS	None	None
5	None	PhCH$_2$CH$_2$	Br	None	None
5a			PTS	Blue (610 nm)	Blue (610 nm)
6	None	–(CH$_2$)$_3$SO$_3^-$		Blue (610 nm)	Blue (610 nm)
7	None	–(CH$_2$)$_4$SO$_3^-$		Blue (610 nm)	Blue (610 nm)
8	–CH=CH–	PhCH$_2$	Br	None	None
8a			PTS	None	Pink (530 nm)
9	–CH=CH–	–(CH$_2$)$_3$SO$_3^-$		Pink (530 nm)	Pink (530 nm)

[a] Irradiation 90 s, 10 cm from 75-W high-pressure mercury lamp.
[b] Figures in parentheses indicate absorption maxima or reflection minima.
[c] PTS = paratoluenesulphonate.

Table 6.13 Half (dark) recovery from the colour developed on viologen crystals with irradiation by UV light[a]

$$R-\overset{+}{N}\diagbndA\diagbnd\overset{+}{N}-R. \quad 2PTS^-$$

Viologen no.	A	R	Half-recovery/min[b] In air	Under vacuum	In PVP matrix
1a	None	PhCH$_2$	1.1	39.1	2.7
5a	None	PhCH$_2$CH$_2$	26.1	2136	8.1
6	None	–(CH$_2$)$_3$SO$_3^{-c}$	509	542	3,698[d]
7	None	–(CH$_2$)$_4$SO$_3^{-c}$	364	383	142[d]
9	–CH=CH–	–(CH$_2$)$_3$SO$_3^{-c}$	238[e]	380[e]	80.6[d,e]

[a] Irradiation 90 s, 10 cm distance, 75-W mercury lamp.
[b] At 610 nm except where indicated.
[c] No PTS.
[d] Turbid films.
[e] At 530 nm.

Table 6.13 shows the decolourization of previously (UV) photoexcited viologen disulphonate crystals (see Table 6.12), both in the presence of air oxygen and under vacuum. Data for the same viologens in PVP matrix are also given for comparison. It is interesting to note that, at the time of writing (1990), only two groups of viologen crystals have been found to be UV-sensitive (Table 6.12), *viz.* benzylviologens bearing *p*-toluenesulphonate anions (**1a, 5a** and **8a** in Table 6.12) and viologens bearing ω-sulphonatoalkyl groups (**6, 7** and **9** in Table 6.12); the sulphonate groups are common to both types.

The shapes of reflection spectra and the position of their maxima for viologen crystals as shown in Figure 6.4 correspond exactly to those of conventional viologen radical cations in solution. However, the electron spin resonance (ESR) data in Figure 6.5 demonstrate that the production of radical cations is associated with the colour formation as seen from colouration/decolouration experiments. Assuming this to be the case, the redox photochromism of viologen crystals can be considered to proceed according to eqn (7).

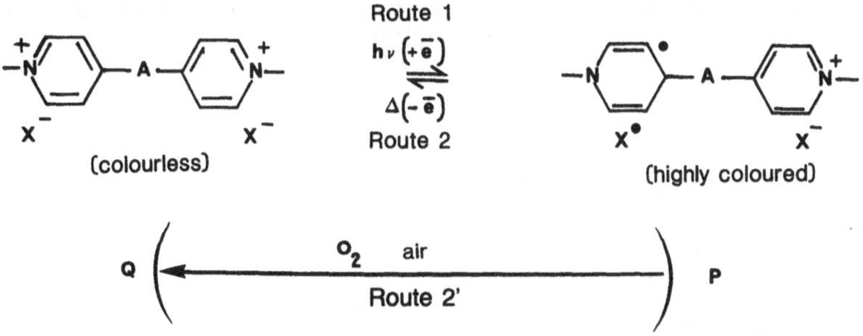

Thus, the viologen dication and perhaps also its counteranion (X^-) are activated by near-UV irradiation to induce the electron transfer between the cation and anion (route 1). The resulting radical cation, the one-electron reduction product of the dication, with simultaneous production of X by the oxidation of X^-, gives rise to a strong colour for P, the one-electron transfer product of the colourless ion pair Q.

The aforementioned mechanism can be illustrated schematically by the energy diagram shown in Figure 6.6. Thus, the potential energy of the ion pair in the ground state (Q) is increased with the absorption of light energy until it surpasses the potential barrier, so that the electron transfer within the activated ion pair may occur to produce P (route 1). Whether route 1 passes through the transition state C (exciplex) or not is not yet clear.

The original dication may be reproduced by the air oxidation of the radical cation (route 2′ in eqn 7). Colour fadings for viologens **1a** and **5a** in Table 6.13 may mainly take place via route 2′.

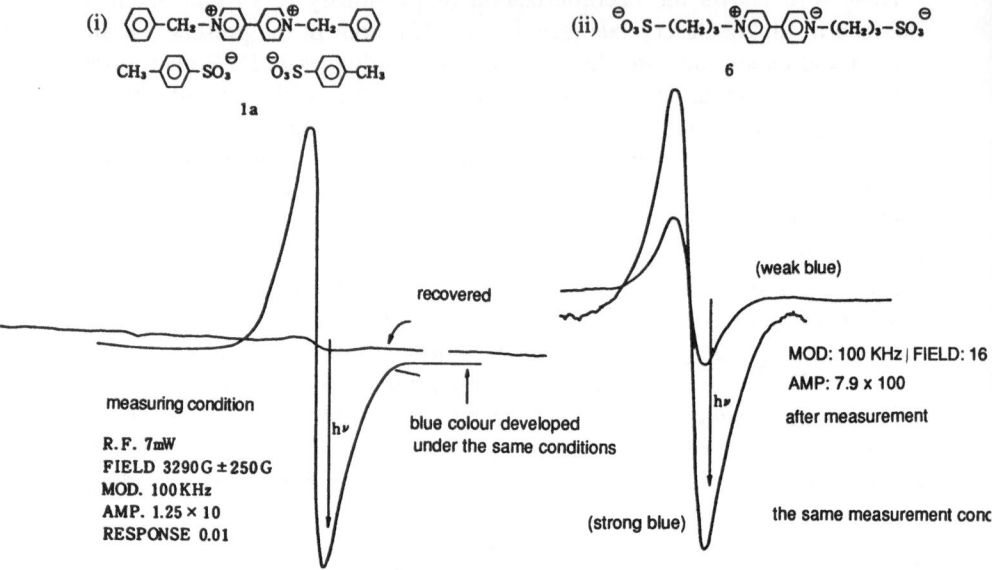

Figure 6.5 Electron spin resonance (ESR) change with the colour development for viologen crystals (see Table 6.12).

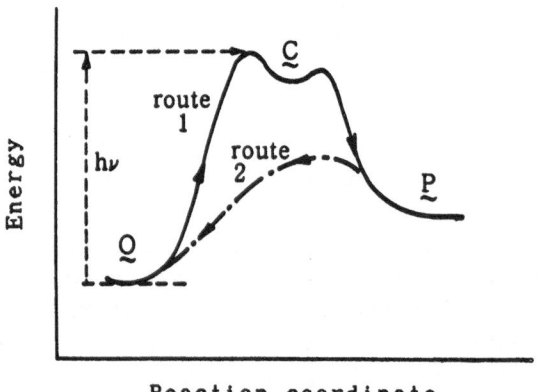

Figure 6.6 The relationship between energy and reaction coordinate for eqn (7).

The back reaction (dark reaction) in eqn (7) (route 2) may also occur thermodynamically. The fact that, in Table 6.13, half-recovery time is little changed between air and vacuum conditions for viologens **6** and **7** may demonstrate that the back reaction via route 2 is the major route of the colour fading. In fact, a blue colour developed for viologen crystal **6** disappears by heating it above 100°C.

It would appear from the data in Tables 6.12 and 6.13 that the thermodynamic stability of the radical ion pair P (which depends on both chemical and crystal structures) predominates in the colour fading as well as in the photocolour development behaviour. The fact that viologen **1a** (Table 6.12) comprising *p*-toluenesulphonate anions indicates a photochromic behaviour entirely different from that of **1b** which comprises the chemically related benzenesulphonate anions (Table 6.12) may illustrate the effect of crystal structure.

Sulphonate anions, if properly arranged, may also contribute to the stability of P through electron delocalization.

Another experimental fact supporting the mechanism of the redox photochromism mentioned above has been found recently by Kamogawa and Ono (1990). Thus, transparent films made of some viologens bearing long alkyl chains alone without any additive exhibit a redox photochromism which is essentially the same as that for viologen crystals, thereby indicating that the photochromism should also follow eqn (7).

6.6 Preparation of polar aprotic viologen copolymers and their photochromic behaviour

This section is concerned with the preparation and redox photochromism of vinyl polymers bearing viologen pendants.

Vinyl monomers bearing the viologen units 1a–c and 2 were synthesized in the same manner as previously described in section 6.2.

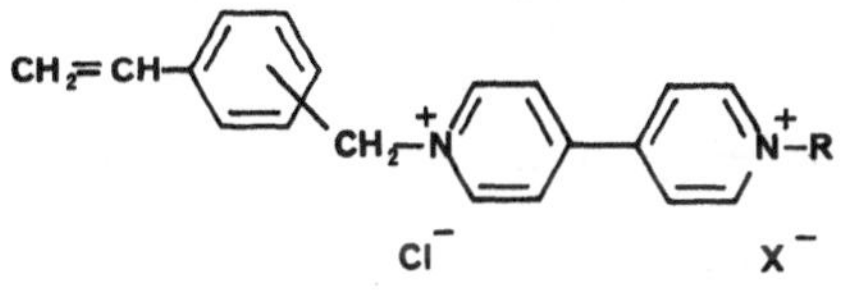

1a : R= *n*-Pr , X = Br
1b : R= PhCH₂ , X= Cl
1c : R= PhCH₂ , X= Br

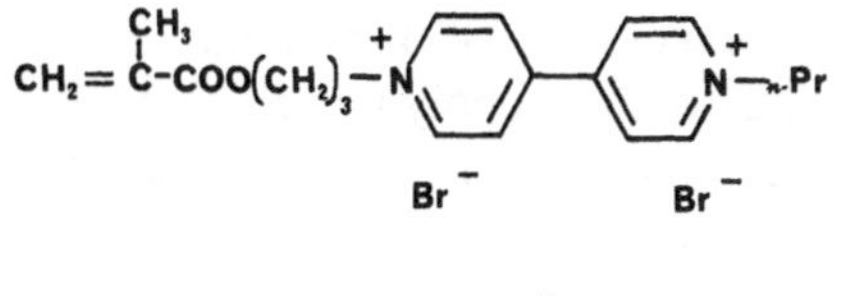

These monomers were copolymerized with various comonomers in water
or ethanol. (Kamogawa and Amemiya, 1985). The results are summarized
in Table 6.14. It is can be seen from this table that copolymerizations with
polar aprotic comonomers such as N,N-dimethylamide (DMA), and protic
ones such as 2-hydroxyethyl arcylate (HEA) and acrylamide (AM) in protic
solvents, e.g. water and ethanol, provide reasonable results. However,
copolymerizations with N-vinyl-2-pyrrolidone (VP) in water afford very low
conversions, accompanied by coloured products insensitive to UV light.

Vinyl polymers bearing the viologen unit as pendant together with the
pendant N-substituted 2-pyrrolidone unit (5, 6 and 9 in eqns 8 and 9) were
prepared by the chemical modification of conventional polymers according
to eqns (8) and (9). Thus, copolymers of chloromethylstyrene (3) and 3-
bromopropyl methacrylate (7) with a large excess of VP were employed
as starting materials. The reactive halogens in 3 and 7 were converted to
the viologen units in the conventional manner.

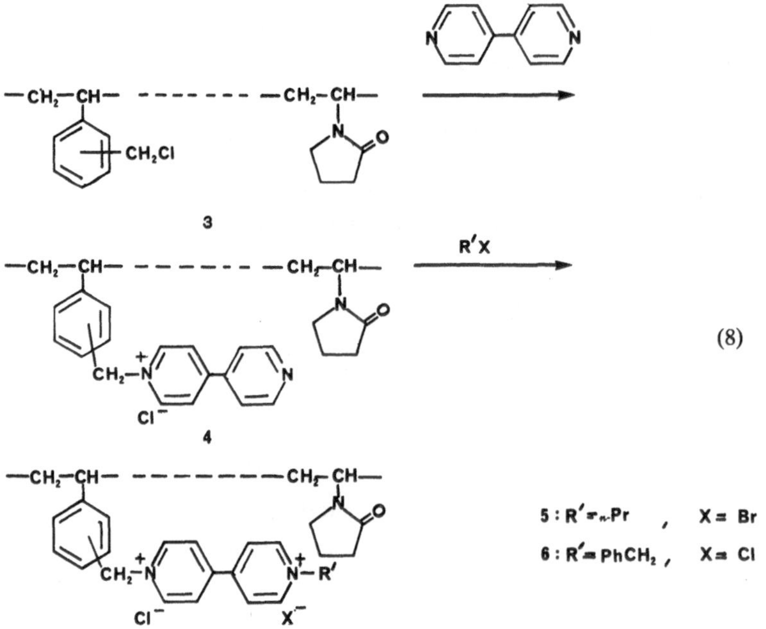

$$(8)$$

5 : R′ = n-Pr , X = Br

6 : R′ = PhCH₂ , X = Cl

In more detail, polymer 3 or 7 is converted to monoquaternary pyridinium
salt (4 or 8) with a large excess of 4,4′-bipyridine. The copolymer 4 or 8
is subsequently converted to the copolymer 5, 6 or 9 bearing both the
viologen unit with a large excess of n-propyl or benzyl halide. DMF is a
suitable solvent for these reaction series.

Copolymers of 1b in Table 6.14 (except those with VP) were subjected
to an anion-exchange reaction using anion-exchange resins to give the

Table 6.14 Copolymerization behaviour of viologen monomers.

Experiment no.	Viologen monomer[b] (g)	Comonomer[b] (g)	Catalyst[c] (g)	Solvent (ml)	Polymerization temperature (C)	Polymerization time (h)	Polymer[a] Conversion (%)	Polymer[a] $[\eta]^d$ (dL/g)
101	1a (0.2)	VP (4)	KPS (0.04)	H$_2$O (4)	90	24	12	—
102	1b (0.1)	VP (2)	K$_2$SO$_3^e$ (2.7)	H$_2$O (14)	50	48	5	—
103	1c (0.1)	VP (2)	30% H$_2$O$_2$ (0.3)	H$_2$O+28% NH$_3$ (4+0.3)	50	72	19	—
104	1a (0.1)	DMA (2)	AIBN (0.05)	EtOH (10)	70	72	57	0.16
105	1b (0.1)	DMA (2)	AIBN (0.05)	EtOH (10)	70	72	67	0.20
106	2 (0.1)	DMA (2)	AIBN (0.05)	EtOH (10)	70	72	76	0.18
107	1a (0.1)	HEA (2)	AIBN (0.05)	EtOH (10)	70	72	52	0.08
108	1b (0.1)	HEA (2)	AIBN (0.05)	EtOH (10)	70	72	71	0.23
109	2 (0.1)	HEA (2)	AIBN (0.05)	EtOH (10)	70	72	81	0.12
110	1a (0.1)	AM (2)	KPS (0.05)	H$_2$O (50)	80	2	95	1.29
111	1b (0.1)	AM (2)	KPS (0.05)	H$_2$O (50)	80	2	95	0.63
112	2 (0.1)	AM (2)	KPS (0.05)	H$_2$O (50)	80	2	98	2.25

[a] Elemental analyses; C 61.10, H 8.80, N 11.70% (104), 69.13, 9.11, 13.36 (105), 58.03, 9.67, 13.41 (106), 52.00, 7.10, 0.70 (107), 57.80, 7.40, 0.90 (108), 54.50, 6.92, 0.83 (109), 54.30, 6.90, 16.60 (110), 57.13, 7.25, 17.73 (111), 52.67, 7.39, 16.35 (112).

[b] VP, N-vinyl-2-pyrrolidone; DMA, N,N-dimethylacrylamide; HEA, 2-hydroxyethyl acrylate; AM, acrylamide.

[c] KPS, potassium persulphate; AIBN, α,α′-azobisisobutyronitrile.

[d] Intrinsic viscosity in water at 26°C.

[e] See Sørenson-Campbell (1968) *Preparative Methods of Polymer Chemistry*, 2nd Ed., Interscience, New York, p. 250.

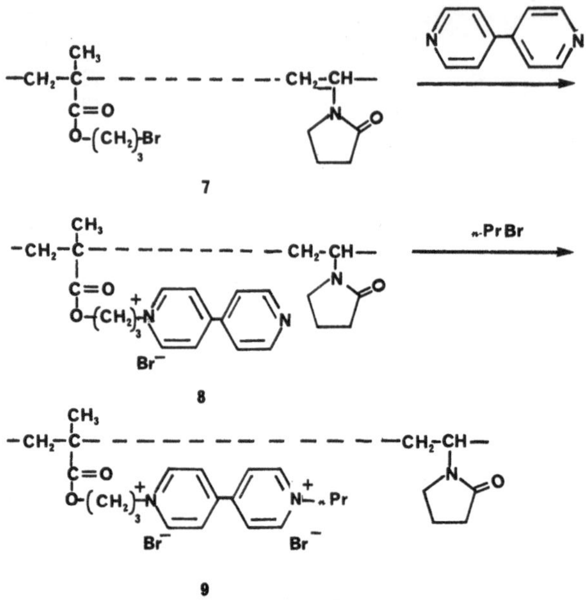

7

(9)

8

9

polymers **10a–d** according to eqn (10). Films of the copolymers thus prepared were made on glass plates by evaporating aqueous or alcoholic polymer solutions at room temperature, as described in section 6.3.

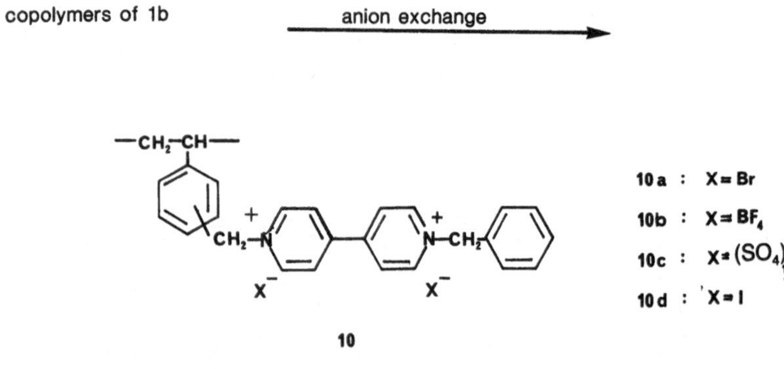

copolymers of **1b** ⟶ anion exchange ⟶

10a : X = Br
10b : X = BF$_4$
10c : X = (SO$_4$)$_{1/2}$
10d : X = I

10

Figure 6.7 indicates the redox photochromic behaviour of the copolymer **6** (see eqn 8) as compared with the solid solution of *N,N'*-dibenzyl-4,4'-bipyridinium dichloride embedded in a PVP film. It is apparent from this figure that the VP copolymers bearing the 4,4'-viologen unit behave in the same manner as their low molecular weight models in a PVP matrix, except that the absorption peak below 600 nm (presumably due to viologen aggregates) which is discernible in the model films hardly appears in the case of the copolymer. This may be a consequence of fixing the viologen unit to the polymer molecular chain, thus preventing aggregation of viologen

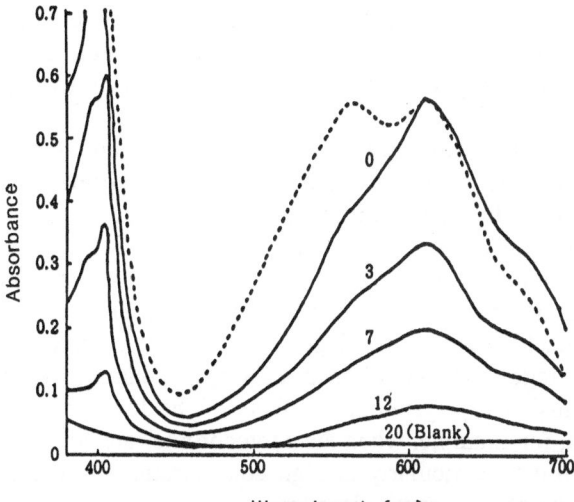

Wave length, (nm)

Figure 6.7   Visible absorption spectra (solid lines) for the film of the copolymer with structure 6 kept at RH 58%, subjected to 60 s irradiation at a distance of 15 cm from a 75-W high-pressure mercury lamp. Figures in the diagram denote times (min) after irradiation. Dotted line: absorption spectrum of an irradiated film of PVP bearing N,N′-dibenzyl-4,4′-bipyridinium dichloride. This spectrum is included to show the shape and has no quantitative significance for absorbance.

unit to the polymer molecular chain, thus preventing aggregation of viologen systems.

The dark recovery of the absorption spectrum (colour fading) is complete in copolymer films, as shown in Figure 6.7.

Table 6.15 summarizes the photochromic behaviour of copolymers **5,6** and **9** from eqns (8 and 9).

Table 6.15 Photochromic behaviour of copolymers with VP.

Polymer[a]	RH (%)	Absorbance maximum (nm)	$\varepsilon_{eq} \times 10^3$ (cm⁻¹)	
			Immediately after irradiation	Recovery (time)
5	84	608	3.2[b]	0 (5 min)
	58	608	3.8[b]	0 (5 min)
	30	608	1.4[b]	0 (25 min)
6	84	608	15.4[b]	0 (20 min)
	58	608	10.8[b]	0 (20 min)
	30	608	11.4[b]	0 (20 min)
9	84	606	2.4[c]	0 (5 min)
	58	606	2.2[c]	0 (5 min)
	30	606	3.7[c]	0 (5 min)

[a] Elemental analysis: C 64.11, H 7.43, N 11.59% (**5**), 66.10, 7.99, 11.05 (**6**), 60.23, 7.11 10.85 (**9**).
[b] 30 s irradiation at the distance of 5 cm from a 75-W mercury lamp.
[c] 90 s irradiation at the distance of 5 cm from a 75-W mercury lamp.

As in the case of solid solutions of low molecular weight viologens in a PVP matrix, the copolymer **6** bearing the 4,4′-dibenzyl unit and two chloride ions exhibits the highest photosensitivity. The photosensitivity increases, as expected, in the order **9** < **5** < **6** (eqns 8 and 9).

The effect of relative humidity (RH) in the atmosphere is not clear, although generally it is considered that higher RHs afford higher water contents in polymer films, so that films may be plasticized and the optimum conditions for the reversible photoreduction of the viologen unit in the polymer may be reached (see Table 6.10), thereby permitting more facile redox photochromic reactions (see eqn 7). Table 6.15 also shows that dark recovery, i.e. bleaching of the developed colours, was complete.

Another advantage of the copolymers bearing pendant viologen units relates to the exclusion of crystallizing effects of the low molecular weight viologen as seen, for example, when the photochromes are simply embedded in PVP matrix films (particularly at high concentration), and the resulting stabilization of the photochromic function.

Table 6.16 indicates the photochromic behaviour of the copolymers with DMA which constitute another polar aprotic polymer system.

Table 6.16 Photochromic behaviour of copolymers with DMA.

Polymer	RH (%)	Absorbance maximum (nm)	$\varepsilon_{eq} \times 10^3$ (cm^{-1})	
			Immediately after irradiation[b]	Recovery (time)
104[c]	84	608	2.4	0.1 (1 h)
	58	608	2.1	0 (1 h)
	30	608	1.4	0 (1 h)
105[c]	84	608	2.2	0.9 (10 min)
	58	608	1.9	0.7 (10 min)
	30	608	2.1	0 (10 min)
10b–DMA[a]	84	608	1.4	0 (5 min)
	58	608	1.7	0 (5 min)
	30	608	0.7	0 (5 min)
10c–DMA[a]	84	608	2.4	0.3 (5 min)
	58	608	1.3	0 (5 min)
	30	608	2.1	0.2 (5 min)
10d–DMA	84	608	0.2	0 (5 min)
	58	608	1.3	0 (5 min)
	30	608	0.5	0 (5 min)
106[c]	84	605	3.7	0 (10 min)
	58	605	3.1	0.6 (10 min)
	30	605	3.7	0 (10 min)

[a] Elemental analysis: C 65.55, H 8.44, N 12.38% (**10b**–DMA), 65.45, 8.76, 12.44 (**10c**–DMA), 63.51, 8.25, 11.61 (**10d**–DMA).
[b] 30 s irradiation at the distance of 5 cm from a 75-W mercury lamp.
[c] Polymers in Table 6.14.

Tables 6.17 and 6.18 show the results obtained for protic copolymer systems prepared with HEA and AM respectively.

Table 6.17 Photochromic behaviour of copolymers with HEA.

Polymer[a]	RH (%)	Absorbance maximum (nm)	$\varepsilon_{eq} \times 10^3$ (cm^{-1}) Immediately after irradiation	Recovery (time)
107[b]	84	608	0.6[c]	0 (1 min)
	58	608	0.6[c]	0 (1 min)
	30	608	0.6[c]	0 (1 min)
108[b]	84	610	0.1[d]	0 (5 min)
	58	610	0.2[d]	0 (5 min)
	30	610	0.2[d]	0 (5 min)
10a–HEA	84	608	0.5[e]	0 (5 min)
	58	608	0.2[e]	0 (5 min)
	30	608	0.7[e]	0.1 (5 min)
10b–HEA	84	608	0.5[e]	0 (5 min)
	58	608	0.3[e]	0 (5 min)
	30	608	0.4[e]	0 (5 min)
10c–HEA	84	609	1.1[c]	0 (10 min)
	58	609	0.7[c]	0 (10 min)
	30	609	1.2[c]	0 (10 min)
109[b]	84	605	0.3[d]	0 (5 min)
	58	605	0.3[d]	0.1 (5 min)
	30	605	0.3[d]	0.1 (5 min)

[a] Elemental analysis: C 56.45, H 6.64. N 0.47% (10a–HEA), 58.44, 6.76, 0.48 (10b–HEA), 52.54, 7.22. 0.82 (10c–HEA).
[b] Polymers in Table 6.14.
[c] 30 s irradiation at the distance of 5 cm from a 75-W mercury lamp.
[d] 90 s irradiation.
[e] 300 s irradiation.

The redox photochromic sensitivity of copolymers with DMA in Table 6.16 appears to be inferior to that of copolymers bearing the N-substituted pyrrolidone unit in Table 6.15. The photochromic function, however, is somewhat discernible, except for copolymers bearing the counteranions BF$_4^-$ (10b–DMA) and I$^-$ (10d–DMA).

Copolymers with HEA comprise many protic hydroxy functions and generally have very low photosensitivities (see Table 6.17). This may be so for the same reason given in section 6.3; thus the viologen counteranion portion in polymer has a high affinity and is stabilized and hence deactivated by the surrounding hydroxy groups. In contrast, the highly coloured radical ion pair portion, even if produced by light according to eqn (7), is less stable because of its decreased hydrophilic character and so would promote the back reaction to the colourless original ion pair.

Table 6.18 Photochromic behaviour of copolymers with AM.

Polymer[a]	RH (%)	Absorbance maximum (nm)	$\varepsilon_{eq} \times 10^3$ (cm^{-1})	
			Immediately after irradiation	Recovery (time)
110[b]	84	610	1.4[c]	0 (1 h)
	58	610	0.9[c]	0 (1 h)
	30	610	0.6[c]	0 (1 h)
111[b]	84	610	0.9[d]	0.5 (5 min)
	58	610	0.4[d]	0 (5 min)
	30	610	0.3[d]	0 (5 min)
10a–AM	84	606	2.1[c]	0 (20 min)
	58	606	1.4[c]	0 (20 min)
	30	606	2.8[c]	0 (20 min)
10b–AM	84	608	1.1[c]	0 (10 min)
	58	608	1.0[c]	0 (10 min)
	30	608	0.6[c]	0.2 (10 min)
10c–AM	84	608	3.9[c]	2.9 (20 min)
	58	608	3.2[c]	4.3 (20 min)
	30	608	3.0[c]	0 (20 min)
10d–AM	84	608	1.1[d]	0 (10 min)
	58	608	1.0[d]	0 (10 min)
	30	608	1.1[d]	0 (10 min)
112[b]	84	608	1.1[d]	0.6 (20 min)
	58	608	0.7[d]	0.3 (20 min)
	30	608	1.1[d]	0.3 (20 min)

[a] Elemental analysis: C 51.95, H 7.95, N 16.95% (**10a**–AM), 55.42, 6.90, 15.98 (**10b**–AM), 51.75, 6.92, 18.11 (**10c**–AM), 50.23, 6.49, 15.57 (**10d**–AM).
[b] Polymers in Table 6.14.
[c] 30 s irradiation at the distance of 5 cm from a 75–W mercury lamp.
[d] 90 s irradiation.

This relationship is just the opposite in polar aprotic polymers such as those in Tables 6.15 and 6.16: the viologen counteranion portion is activated because of its low affinity for the polar aprotic N-substituted pyrrolidone or N,N-dimethylamide portion, whereas the radical ion pair portion is more stable as a result of the increased affinity for the polar aprotic portion.

It is interesting to note, however, that some copolymers with AM such as **10a**-AM and **10c**-AM in Table 6.18 are almost equal in photosensitivity to those with DMA in Table 6.16, in spite of the protic nature of the former. This might suggest that the affinity of the amide portions for the viologen ion pair portions in these copolymers is not as large as in the hydroxy portion in copolymers in Table 6.17.

Summarizing the results of the redox photochromic behaviour obtained with the polymer films in section 6.6, it can be said generally that both the sensitivity to UV light and the recovery rate in the dark for polar aprotic viologen polymers compare favourably with those for the corresponding low molecular weight viologens embedded in polymer matrix films, especially when embedded in PVP, while simultaneously maintaining all the advantages of copolymeric systems.

6.7 Photochromism of N-aryl viologens

As indicated in Table 6.19 (Kamogawa and Satoh, 1988) the N-aryl viologens to be described in this section are N,N'-disubstituted 4,4'-bipyridinium salts, in which two aryl groups are conjugated at the N,N' positions with a 4,4'-bipyridinium ring.

Table 6.19 Photochromic N-aryl viologens.

N-aryl viologens (IV) were synthesized according to eqn (11). Step I involves an additional reaction of the chlorobenzene (II) (substituted with 2,4-dinitro groups for activation) to 4,4'-bipyridine (I), thus a mixture of I and II is heated to reflux in acetonitrile for 72 h to produce the viologen III in almost quantitative yield.

In step II, the activated 2,4-dinitrophenyl substituents in III are replaced by the aryl groups indicated in Table 6.19 using the corresponding aryl amines (ArNH$_2$). The reaction is carried out by adding an appropriate aryl amine to a solution of III in aqueous ethanol, followed by evaporation of the solvent and subsequent boiling with water for 24 h to give IV in 22–99% yield.

In the case of aryl viologens 8 and 9 (Table 6.19), the mixture of III and

$$(11)$$

aryl amine is stirred at 80°C for 24 h prior to boiling with water, because of the low reactivity of the amine. Yields of 64 and 15% were obtained for viologens **8** and **9** respectively. The synthesis of N-aryl viologens bearing the p-nitrophenyl N-substituent failed, indicating that the more electron-withdrawing the aryl amines employed for the exchange reaction in eqn (11), the less likely is the chance that the reaction will take place.

Unsymmetrical N-aryl viologens, i.e. those in which the N- and N'-aryl substituents differ from one another, can be prepared in a manner similar to that outlined above, by first quaternizing one of the 4-pyridyl groups of **I**, then applying the exchange reaction with an aryl amine, then applying the same reaction sequence to the resulting N-arylation product with a second aryl amine (see section 6.2 and eqn (11)). Ethanol is used as the solvent throughout the process.

IV

The counterion Cl⁻ of the viologen **IV** can be exchanged readily with another one such as bromide and p-toluenesulphonate using an anion-exchange resin according to conventional procedures. Transparent, almost colourless, films of the N-aryl viologen (**IV**) embedded in the PVP matrix can be prepared on a glass plate by evaporating a solution of **IV** and PVP in water as outlined in section 6.2.

Typical visible absorption spectra showing the colours developed for such films when subjected to near-UV irradiation are given in Figure 6.8. The absorption maxima for the colours thus developed are further summarized in Table 6.20.

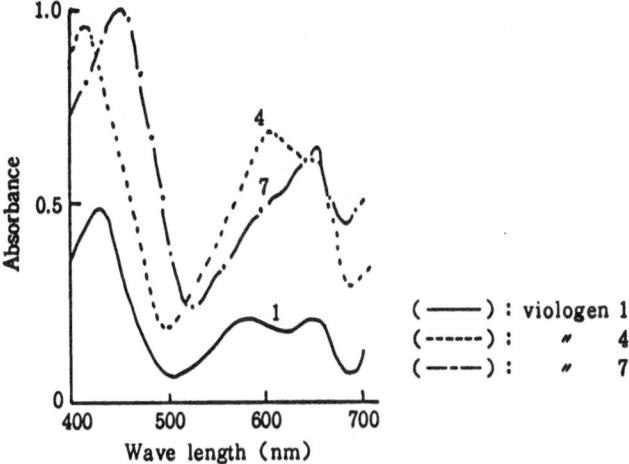

Figure 6.8 Visible absorption spectra indicating the colours developed in various viologens embedded in PVP matrix films with irradiation of UV light. ——, viologen no. 1; - - - -, viologen no. 4; — - —, viologen no. 7.

It is noteworthy from these data that the absorption spectra for the radical cations of N-aryl viologens (with the exception of the one bearing the o-methylphenyl group in Table 6.20) are red-shifted as compared with those from the conventional 4,4'-viologens such as benzyl viologen. The red shifts amount to 15–40 nm for the absorption peak around 410 nm and to 35–40 nm for the peak around 610 nm and may be a consequence of the extension of the delocalization of the pi system to include the· radical cation. The irradiated film containing the viologen **IV** is intense green in colour. The fact that red shifts do not exist in the case of the viologen bearing the o-Me phenyl N-substituent is considered to be a result of the '$ortho$ effect', which might hinder the participation of the aryl group to the delocalization of pi electrons.

Table 6.20 Effect of substituent on visible absorption maxima of photocoloured aryl viologens in PVP matrix.

	Absorption maximum (nm)	
Substituent	Near-UV	Near-IR
None	425	645
p-Me	430	645
m-Me	430	650
o-Me	410	605
p-F	440	650
p-Cl	450	650
p-OCH$_3$	450	650
(Benzyl viologen)	(410)	(610)

As shown in Table 6.21, the sensitivities to UV light of N-aryl viologens in the PVP matrix film are 3 to 30-fold higher than those of benzyl viologen (N,N'-dibenzyl-4,4'-bipyridinium dichloride). As expected, the sensitivity increases with the electron-withdrawing ability of the N-substituted aryl group, corresponding closely with the Hammett sigma values, to the extent that for viologens **8** and **9** (Table 6.21) bearing m-nitro and p-cyanophenyl N-substituents, the films prepared from almost colourless aqueous solutions develop colours spontaneously and colour fading does not occur.

Table 6.21 Absorbance of PVP film containing aryl viologen developed with irradiation by UV light.

Viologen[a]	Absorbance at absorbance maximum reduced (645–650 nm) (mm^{-1})	Half recovery[b] in air (min)	σ^c
1	5, 1	20	0.00
2	5, 1	30	−0.170
3	8.2	60	−0.069
4	19.5^d	30	—
5	1, 8	60	−0.268
6	32.0	240	+0.062
7	41.9	>120	+ 0.227
8	e	—	+0.710
9	e	—	+0.660
Benzyl viologen	1.5	—	—

Note that irradiation time was 10 s.
[a] See Table 6.19.
[b] RH, 58%.
[c] Hammett σ values.
[d] 605 nm peak.
[e] Spontaneous colour development.

It should be noted that, in this case, the absorptions between 500 and 600 nm, which are attributable to the aggregates of the radical ion pairs produced from viologen ion pair molecules, are generally more significant compared with the blue colour which results from conventional radical cations, as shown in Figure 6.8.

6.8 Viologens developing infrared absorption peaks (near 800 nm) by redox photochromism

Recently, compounds having absorption in the near-IR region, particularly around 800 nm, have been attracting much interest from a device standpoint because the emission wavelengths of semiconductor lasers usually cover this wavelength range (see Chapter 1). This being the case, photochromism in the near-IR region is an active area of research. Thus, for example, it was reported by Japanese workers (Arakawa *et al.*, 1985) that spirothiopyrans containing the thioether structure showed absorption after irradiation which extended up to the near-IR region, although their maxima were mostly in the visible region.

In the viologen case, compound **2** in Figure 6.9 was first found to develop the absorption spectrum having a broad maximum around 800 nm as a result of UV excitation when embedded in a PVP matrix film (Kamogawa and Sugiyama, 1985). The characteristics of such viologens are given in Table 6.22 and Figure 6.10.

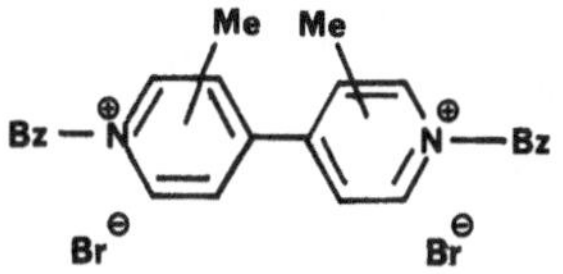

$$2,2'-\text{Me} \quad 1$$
$$3,3'-\text{Me} \quad 2$$

Figure 6.9 Viologens capable of photogenerating species with near-infrared absorption.

Table 6.22 Photochromic behaviour of compounds 1 and 2 in PVP matrix.

Viologen[e]	Irradiation[a] time (s^{-1})	Absorption peak (nm)	Absorbance[b] (mm)$^{-1}$	Half-recovery[c] (min)
Benzyl viologen dibromide	30	610 570[d]	4.82 —	30 —
1	30	605	4.42	20
2	60	775	2.65	30

[a] At a distance of 10 cm from a 75W mercury lamp.
[b] Reduced to 1 mm of film thickness.
[c] Recoveries in the dark were complete for all films.
[d] Peak for aggregate.
[e] See Figure 6–9.

In order to synthesize viologens **1** or **2**, 2- or 3-picoline is first dimerized with sodium metal to provide 2,2'- or 3,3'-dimethyl-4,4'-bipyridine, which is converted to the target viologen using benzyl bromide in the conventional manner. As can be seen from the absorption spectrum II in Figure 6.10, together with the results in Table 6.22, viologen **1** with 2,2'-dimethyl substituents displays the usual photochromic behaviour, except that the absorptions between 500 and 600 nm (attributable to the aggregation products of radical cations) which are pronounced for normal viologen radical cations, e.g. spectrum I, are barely noticeable in the spectra. This may be a consequence of the methyl substituents, which through steric interaction prevent aggregation. However, in the case of viologen **2** embedded in a PVP matrix film, the visible absorption spectrum for the irradiated film is entirely different from those of conventional viologen radical cations, as can be seen in Figure 6.10: absorption spectrum III for the irradiated film containing viologen 2 shows a broad absorption with a maximum at 775 nm.

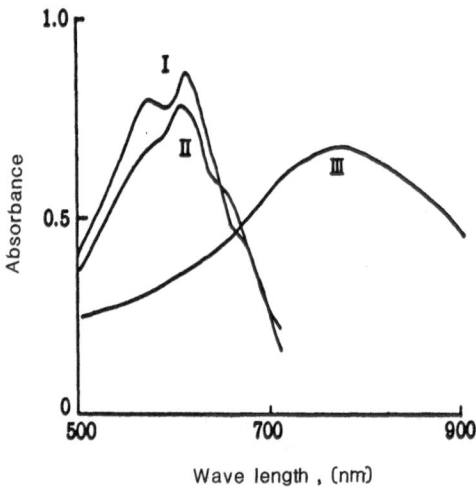

Figure 6.10 Absorption spectra obtained for viologens in PVP matrix films under exposure to UV light. I, benzyl viologen dibromide; II, viologen no. 1 in Table 6.22; III, viologen no. 2. Irradiation of films was carried out at a distance of 10 cm from a 75-W high-pressure mercury lamp. Irradiation times were 30, 30, and 60 s for I, II, and III respectively.

This striking difference in absorption spectra, which makes the photochromism involving the spectral change in the near-IR possible, appears to be caused by the steric hindrance induced by methyl groups located at the 3,3′ positions of the 4,4′-bipyridinium ring, which hinders the usual delocalization of pi electrons of viologen radical cations.

Other types of viologen which exhibit redox photochromism in the near IR are as follows (Kamogawa and Nanasawa, 1988).

Viologens belonging to group 1 are synthesized either by quaternizing the *trans* form of **I** or by preparing the compound with R₃ as the 3-substituent for **I** by the dehydrative condensation of 4-pyridine aldehyde with the compound having R₃ as the 3-substituent of 4-picoline, followed by quaternization.

Viologens belonging to group 2 are synthesized either by quaternizing 2,2'-bipyridine or its 4,4'-disubstitution products with ethylene dibromide (Murase, 1956; Maerker and Case, 1958; Sprintschnik *et al.*, 1977).

Viologens from groups 1 and 2 were incorporated into the PVP matrix and subsequently the films were sandwiched tightly between two glass plates to exclude the effect of atmospheric oxygen. The sandwiched films (<0.1 mm thick) thus prepared were then subjected to UV irradiation.

Figure 6.11 indicates the absorption spectrum of the film bearing a typical viologen from group 1 immediately after UV irradiation, compared with that for conventional benzyl viologen. It can be seen from this figure that the absorption spectrum corresponding with the radical cation in this case has a visible absorption around 500 nm, thereby giving rise to a red-coloured film (dotted curve). Note the large absorption peak around 800 nm, which coincides with the output wavelength of typical semiconductor lasers. Presumably the absorption peaks around 400 and 600 nm, characteristic of the conventional 4,4'-viologen radical cation (solid line), are red-shifted to about 500 and 800 nm respectively because of the promotion of delocalization of pi electrons via the intervening –CH=CH– system in these molecules.

Another feature of the viologens from group 1 is the rapid dark recovery of the absorption spectra of irradiated films in spite of the absence of atmospheric oxygen, as shown in Figure 6.12. This is quite different from the

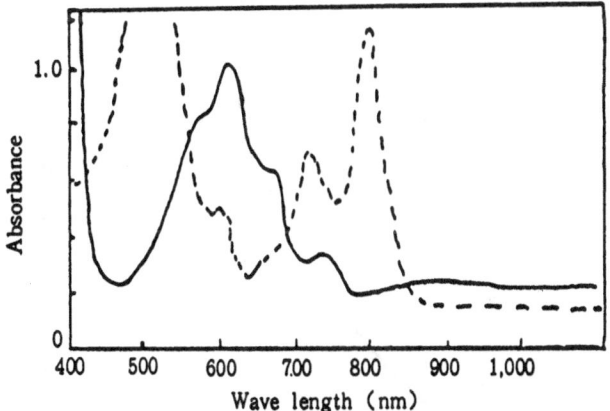

Figure 6.11 Absorption spectra under illumination: ——, 1,1'-dibenzyl-4,4'-bipyridinium dichloride; – – – –, trans-4,4'–vinylenebis(1-benzylpyridinium) dichloride, both embedded in PVP matrix films.

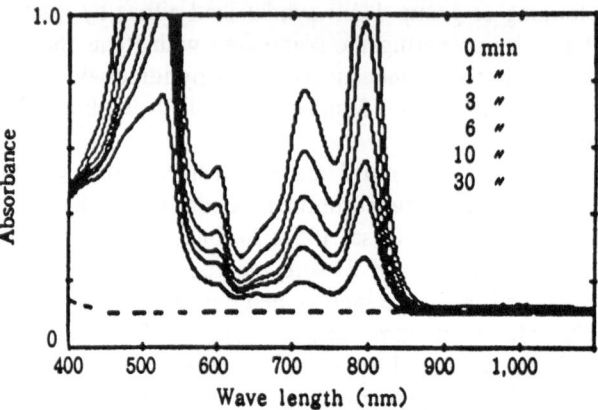

Figure 6.12 Spontaneous bleaching of the absorption spectrum developed by light for a PVP film bearing *trans*- 4,4′-vinylene bis (1-benzylpyridinium) dichloride (belonging to group 1). Figures in the diagram denote bleaching time at 25°C.

behaviour of PVP films bearing 4,4′-viologens such as benzyl viologen, in which the developed blue colour persists for more than 1 year. This would tend to suggest that the thermodynamic stability of the red-coloured radical cation — anion pairs produced according to eqn (12) is inferior to that of 4,4′-viologens, so much so that when irradiation is stopped the resulting dark equilibrium is driven to the left (viologen ion pairs) of eqn (12).

Figure 6.13 shows the result of UV irradiation on the glass-sandwiched

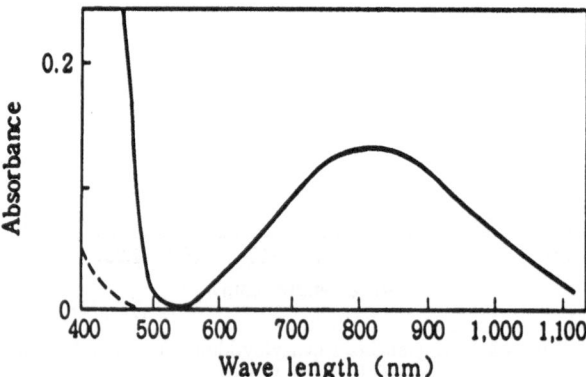

Figure 6.13 Typical absorption spectra for 1,1′-ethylene-2,2′-bipyridinium dibromide (belonging to group 2) embedded in PVP matrix film. ———, upon irradiation; - - - -, before irradiation.

PVP film bearing a typical viologen from group 2. A broad absorption peak is again observed around 800 nm. Although these absorptions can be maintained for fairly long times when air is excluded, the photocolouration and subsequent dark decolourization can be cycled many times, even when the films are exposed to air oxygen as in the case of 4,4′-viologens.

Table 6.23 Near-IR absorption peak developed for compounds **1** and **2** in PVP matrix with irradiation by light[a].

Sample no.	Compound	R₁	R₂	R₃	Visible (nm)	Near-IR (nm)
					Absorbance peak developed by light	
1	**1**[b]	$PhCH_2$	$PhCH_2$	H	530	790
2	**1**[b]	$PhCH_2$	$PhCH_2$	Me	540	790
3	**1**[b]	$PhCH_2$	$PhCH_2$	Et	530	790
4	**1**[b]	$p\text{-MePhCH}_2$	$PhCH_2$	H	530	790
5	**1**[b]	$p\text{-ClPhCH}_2$	$PhCH_2$	H	530	790
6	**1**[b]	$n\text{-C}_3\text{H}_7$	$n\text{-C}_3\text{H}_7$	H	None	None
11	**2**[c]	H	—	—	—	800
12	**2**[c]	Me	—	—	—	730
13	**2**[c]	$CONMe_2$	—	—	—	870
14	**2**[c]	OEt	—	—	—	680
15	**2**[c]	Br	—	—	—	None

[a] For glass-sandwiched film with irradiation from a 75-W mercury lamp.
[b] Irradiation: 10 s at 10 cm from the lamp.
[c] Irradiation: 120 s at 5 cm from the lamp.

Table 6.23 summarizes the results of the redox photochromic behaviour of the viologens from groups 1 and 2. Viologens of group 1 show high sensitivities to UV light compared with those of group 2 viologens. Note the fact that the majority of these viologens show photoinduced absorption peaks in the near-IR region (700–900 nm).

6.9 Comparison of various viologens and photochromic reaction mechanisms in the poly(N-vinyl-2-pyrrolidone) (PVP) matrix

In this section the photochromic behaviour of the three general types of viologens so far described is compared. The three types are summarized by the structures given below:

1	R	$PhCH_2$,	A none
2	R	Ph,	A none
3	R	$PhCH_2$,	A $-CH=CH-$

Visible absorption spectra developed for these viologens embedded in the PVP matrix following UV irradiation are shown in Figure 6.14. The absorption peaks for the radical cations thus produced, except for those aggregates, are situated at about 400 and 610 nm, 425 and 645 nm, and 530 and 790 nm for viologens **1**, **2** and **3**, which give rise to the visible colours blue, green and red respectively.

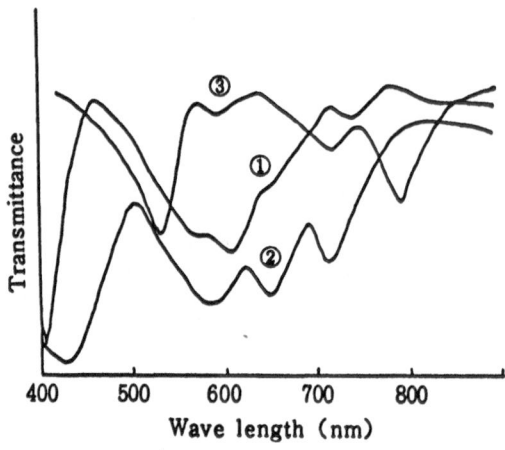

Figure 6.14 Comparison of the absorption spectra of PVP films bearing three kinds of viologen upon UV irradiation.

The sensitivity to UV light is in the order **2, 3 > 1**. The thermodynamic stabilities of the photogenerated species for the glass-sandwiched PVP films comprising these viologens are described in Figure 6.15. It can be seen from this figure that the blue colour from viologen **1** is the most stable (unchanged for more than 1 year when air is completely excluded), the green colour from **2** is intermediate and the red colour from **3** is the least stable of the three and displays bleaching.

Considering these results together with those describing the viologen crystals (section 6.5), it can be seen that the redox photochromism of viologens in PVP matrix films generally proceeds according to the scheme depicted in eqn (13). Thus the probability that viologens exist in the form of ion pair **I** in a PVP matrix film appears also to be high, given their low solubilities in MP as well as the low affinity of polar aprotic PVP for the counteranions X_2^{2-}. Accordingly V^{2+} and X^- in **I** are considered to be closely associated with the 'naked anion X^-' and activated by polar aprotic solvents such as MP so that the ion pair **I** may be in the high-energy state.

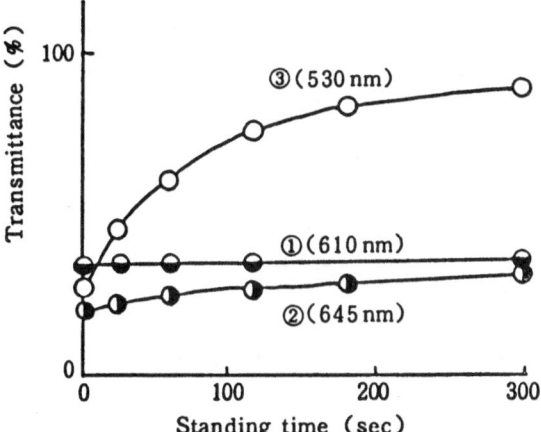

Figure 6.15 Comparison of the stabilities of the absorption peaks developed upon irradiation.

Considering the low solubility of viologens in media containing the pyr-rolidone moiety, the affinity of PVP for V^{2+} may be much less than that for DMF, DMSO and the like, so that **I** might represent the most activated, highest energy state, which is closely related to that in the crystalline state.

The presence of water in PVP is sure to stabilize **I** by hydration, thereby lowering the level of energy from **I** to **I′** (partial hydration product of **I**), as shown in Figure 6.16. This may be understood from the fact that viologens in MP without added water develop colours spontaneously and, with addition of water, it is necessary to photoirradiate to develop colour. However, in water alone, colour is no longer developed even with the application of light energy (see Table 6.11). Thus, the ion pair **I–I′** activated to a high-energy state in the PVP matrix could readily surmount the activation energy barrier by irradiation, as illustrated in Figure 6.16 via an exciplex **II–II′** (see eqn (13)) to the radical ion pair **III–III′** by intramolecular electron transfer. (Note that **II′** and **III′** also represent partial hydration products of **II** and **III** respectively.)

$$V^{2+}X_2^{2-} \xrightarrow{h\nu} \left[V^{2+}X_2^{2-}\right]^* \longrightarrow V^{+\bullet}X_2^{-\bullet}$$

| I | II | III | (13) |
| colourless ion pair | Exciplex | highly coloured radical ion pair | |

Route 1 ←————————

Route 2 O_2

In the case of viologens embedded in PVP matrices, moisture is usually present to a level of several per cent depending upon the ambient RH so that the ion pair and its exciplex are somewhat hydrated (**I′** and **II′**) with a potential barrier somewhat higher than that of **I**, as shown in Figure 6.16.

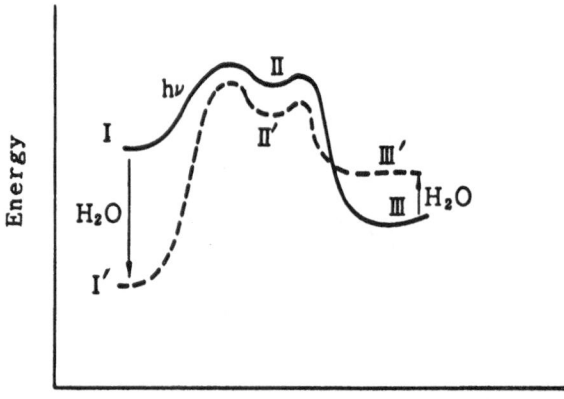

Reaction coordinate

Figure 6.16 Relationship between energy and reaction coordinate indicating redox photochromism of viologen in PVP matrix film.

The presence of a small amount of moisture in a PVP matrix might prevent spontaneous colour development of viologen at room temperature because of this higher potential barrier and thus may subsequently require the application of light energy to promote colour development.

As can also be seen from Figure 6.16, photoproduct **III** has a poor affinity for water but may have a favourable affinity for the PVP matrix in contrast to **I**, so that a low energy level can be attained by stabilization. The presence of moisture in PVP films may destabilize **III** towards **III′** with a higher energy level, in distinct contrast to the relationship between **I** and **I′**. This stabilization of **III–III′** due to a favourable affinity, together with the activation of **I–I′** for the opposite reason, should play an important role in the stability of the colour developed by light for viologens embedded in PVP matrices and can in some instances produce a stable colour for periods of 1 year or more.

On the other hand, the dark reaction (**III–III′** → **I–I′**), i.e. the bleaching of the developed colour, may take place either by the thermodynamic electron transfer from V$_\bullet^+$ to X$_{\bullet 2}^-$ within the more or less hydrated radical ion pair (**III–III′**) (route 1 in eqn (13)) or, when atmospheric oxygen is present, by air oxidation of V$^+$ back to V^{2+} (route 2 in eqn (13)). The relative extent to which routes 1 and 2 predominate in the back reactions in eqn (13) depends upon a variety of factors, such as the chemical structure of viologen, condition of the matrix film, and the presence or absence of air, etc. The above discussion relating to viologen solid solutions in matrix films made up of polar aprotic polymers such as PVP is also applicable to films composed of the copolymers described in section 6.6.

6.10 Photomemories based on viologen photochromism

Given the foregoing discussion, the application of viologen photochromism to 'photomemory' type outlets appears to be feasible. The photomemory function using both polar aprotic copolymers described in section 6.6 and a PVP matrix for low molecular weight viologens is discussed here (Kamogawa *et al.*, 1989). Copolymers and the corresponding low molecular weight viologens employed for this study are indicated by structures **1** and **2** respectively, where the combinations of R and X are given in Table 6.24.

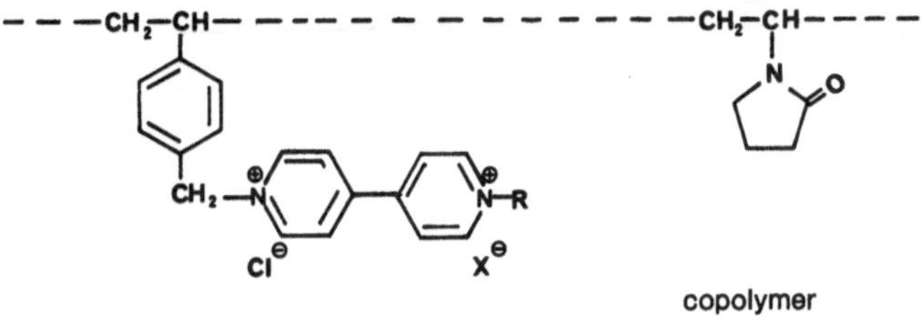

copolymer

1

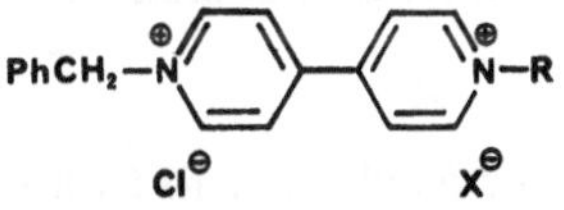

2

Table 6.24 R and X in copolymer **1** and in model compounds **2**.

No.	R	X
1	PhCH$_2$–	Cl
2	PhCH$_2$–	Br
3	C$_3$H$_7$–	Br
4	*p*-MeOPhCH$_2$–	Cl
5	*p*-MePhCH$_2$–	Cl
6	*p*-MePhCH$_2$–	Br
7	*p*-ClPhCH$_2$–	Cl

Films of the copolymers and the PVP films bearing the corresponding viologen **2**, *ca* 0.1 mm thick, were prepared by spreading aqueous solutions over the inner walls of a quartz spectrophotometer cell followed by evaporation of water and subsequent sealing. The cell atmosphere could be alternated between air and nitrogen. UV irradiation was carried out by placing the film in the cell at the distance of 10 cm from a 75-W high-pressure mercury arc lamp for 30 s. The results obtained for typical polymers and the corresponding low molecular weight viologen–polymer solid solutions are given in Figure 6.17 and Table 6.25. From these data it can be seen that, when the cell is filled with nitrogen, absorbances at λ_{max} (610 nm) for copolymer films immediately after irradiation are maintained at usually an approximate 80% level after 12 h storage in the dark, whereas those of the corresponding low molecular weight viologens embedded in PVP reach 90%.

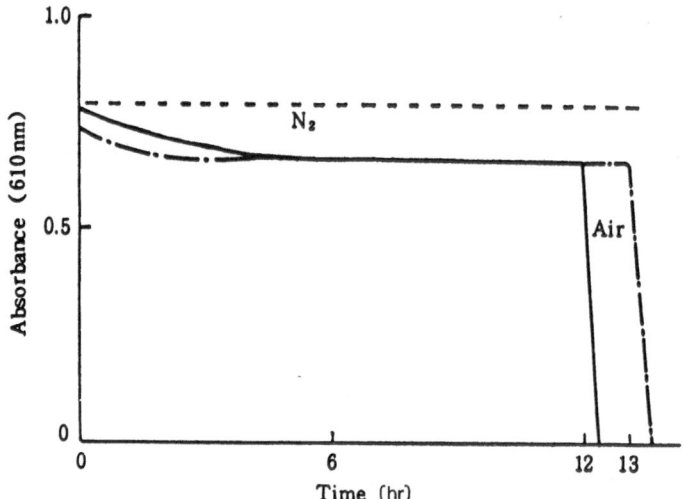

Figure 6.17 Absorption (at 610 nm) change as a function of time for the colour developed by the copolymer with structure **2** (Table 6.24) compared with that for the corresponding viologen in a PVP matrix. ——, copolymer **2**; —— · ——, viologen in PVP matrix; - - - -, copolymer **2** sandwiched between two glass plates.

However, when air is introduced in place of nitrogen the absorbances decrease rapidly and recover those of the original unirradiated state within about 30 min.

PVP films bearing low molecular weight viologens appear preferable, probably because of their more stable radical ion pairs (**III–II** in eqn (13)) than those from the copolymers. The decrease in absorption with time under nitrogen may be a consequence of dissolved oxygen in the original films. However, when glass-sandwiched films are employed, the developed colours are maintained for a long time, as previously discussed.

Table 6.25 Comparison of the relative decay time under nitrogen of colour developed for the copolymer subjected to 30 s irradiation with that for viologen in the matrix.

Polymer no.[c]	A_t/A_o^a	
	Copolymer	Viologen in the matrix
1	0.60 (4)[b]	0.89 (12)
2	0.81 (12)	0.89 (13)
3	0.85 (3)	0.93 (3)
		0.82 (12)
4	0.82 (3)	0.89 (3)
		0.77 (12)
5	0.87 (3)	0.85 (3)
	0.81 (5)	0.49 (15)
6	0.90 (3)	0.94 (18)
7	0.88 (3)	—

[a] A_0 and A_t represent absorbance immediately after irradiation and that after time t under nitrogen respectively.
[b] Figures in parentheses denote time t (h).
[c] See Table 6.24.

This being the case, it is considered conceptually possible that an erasable photomemory using the absorbances at 610 nm for these films could be made by first applying UV light under an nitrogen atmosphere to develop stable colours (photomemory) then providing an erasing function by introducing oxygen. Various other possibilities for erasing colour once developed also exist; these include the application of heat for crystals and related films and also the use of electrochemical methods which may exploit electrochromic phenomena.

References

Arakawa, S., Kondo, H. and Seto, J. (1985) *Chem. Lett*, 1805.
Johns, G. and Matthew, B.Z. (1986) *J. Org. Chem.* **51**, 947.
Johnson, C.S. and Gutowsky, (1963) *J. Chem. Phys.* **39**, 58.
Kamogawa, H., Mizuno, H., Todo, Y. and Nanasawa, M. (1979) *J. Polym. Sci. Polym. Chem. Edn.* **17**, 3149.
Kamogawa, H. and Amemiya, S. (1984) *Rep. Asahi Glass Found. Ind. Technol.* **44**, 42.
Kamogawa, H. Amemiya, S. (1985) *J. Polm. Sci. Polym. Chem. Ed.* **23**, 2413.
Kamogawa, H. and Nanasawa, M. (1988) *Chem. Lett.* 373.
Kamogawa, H. and Satoh, S. (1988) *J. Polym. Sci. Polym. Chem. Ed.* **26**, 653.
Kamogawa, H. and Sugiyama, M. (1985) *Bull. Chem. Soc. Jpn.* **58**, 2443.
Kamogawa, H. and Suzuki, T. (1985) *J. Chem. Soc. Chem Commun.*, 525.
Kamogawa, H. and Suzuki, T. (1987) *Bull. Chem. Soc. Jpn.* **60**, 794.
Kamogawa, H., Masui, T. and Nanasawa, M. (1980) *Chem. Lett.*, 1145.
Kamogawa, H., Masui, T. and Amemiya, S. (1984) *J. Polym. Sci. Polym. Chem. Ed.* **22**, 383.
Kamogawa, H., Kikushima, K. and Nanasawa, M. (1989) *J. Polym. Sci. Polym. Chem. Ed.* **27**, 393.
Kamogawa, H. and Ono, T. (1990) *59th Spring Annual Meeting of the Chemistry Society of Japan, Yokohama, Japan,* 4E104.
Kohjiya, S., Hashimoto, T., Yamashita, S. and Irie, M. (1985) *Chem. Lett.*, 1497.

Maerker, G. and Case, F.H. (1958) *J. Am. Chem. Soc.* **80**, 2745.

Murase, I. (1956) *Nippon Kagaku Zasshi* **77**, 682.

Nambu, Y., Yamamoto, K. and Endo, T. (1986) *J. Chem. Soc. Chem. Commun.*, 574.

Nambu, Y., Yamamoto, K. and Endo, T. (1989) *Macromolecules* **22**, 3530.

Schoot, C.J., Ponjee, J.J., Van Dam, H,T., Van Doorn, R.A. and Bolwijn, P.T. (1973) *Appl. Phys. Lett.* **23**, 64.

Sprintschnik, G., Sprintschnik, H,W., Kirsch, P.P. and Whitten, D.G. (1977) *J. Am. Chem. Soc.* **99**, 4947.

Takuma, K., Kajiwara, M. and Matsuo, T. (1977) *Chem Lett.*, 1199.

Tundo, P., Kippenberger, D.J., Politi, M.J., Klahn, P. and Fendler, J.H. (1982) *J. Am. Chem. Soc.* **104**, 5352.

Index